A GUIDE TO
Home and Garden Pests

A GUIDE TO
Home and Garden Pests
HOW TO IDENTIFY AND ELIMINATE THEM SAFELY

CHARLES KINGSLEY LEVY

Boston University

THE STEPHEN GREENE PRESS
Lexington, Massachusetts

Copyright © The Stephen Greene Press, Inc., 1985
All rights reserved

First published in 1985 by The Stephen Green Press, Inc.
Published simultaneously in Canada by
Penguin Books Canada Limited
Distributed by Viking Penguin Inc., 40 West 23rd Street,
New York, New York 10010

Illustrations on pages 5–43, 68, 148, 158 by Margaret Huong Primack
Illustrations on pages 52–66, 70–76, 80–93 by Laszlo Meszoly
Illustrations on pages 79, 94–142, 149–157, 161–167 provided by the
USDA

LIBRARY OF CONGRESS CATALOGING IN PUBLICATION DATA
Levy, Charles K., 1924–
   A guide to home and garden pests.
   Bibliography: p.
   Includes index.
   1. Household pests.   I. Title.
TX325.L48   1985   648'.7   84-24664
ISBN 0-8289-0546-0

Printed in the United States of America by
The Maple Press Company, York, Pennsylvania
Designed by Irving Perkins Associates

Contents

Acknowledgments

A number of people contributed to this book in a variety of ways and I would like to acknowledge their contributions: Dr. James Traniello for his review of the insect sections, Dr. Stuart Duncan and Dr. Fred Wasserman for their review of the birds and Dr. Thomas Kunz for his contributions to the mammal section. Margaret Huong Primack contributed the majority of the art, particularly in the mammal section, and, in the midst of her artistic efforts, also gave birth to her first child. John Kearney was extremely helpful and creative in the preparation of the manuscript. Some of the art, particularly the representations of the insect pests, came from publications of the U.S. Department of Agriculture. Other drawings were rendered by L. Laszlo Meszoly. I would also like to give special acknowledgement to the dean of Pest Control Studies, Dr. Arnold Mallis of Pennsylvania State University. His thorough *Handbook of Pest Control* is really the magnum opus in the field and served as a most valued reference. Several publications of the U.S. Department of Agriculture were also very helpful.

Many people asked why I didn't include children in a book on pests, but in my case I find this totally inappropriate, since my three children, Brett, Adam and Alison, have been nothing but a pleasure to me, and I dedicate this book to them.

Introduction

While our knowledge about our early human ancestors is far from complete, we can be reasonably certain that they were plagued by a variety of animals in their homes. The problem was exacerbated once humans had mastered the use of fire to keep their homes warm. Furthermore, humans stored a variety of foodstuffs and were then, as now, rather careless housekeepers. Thus, human habitations provided both warmth and sustenance for these animal pests. Although we humans with our large thinking brains consistently improved our life quality, these highly adaptable pest species kept pace. Thus, to our consternation, we are forced to deal with a variety of organisms that intrude on our space, sting and bite us, damage our timbers, furniture and papers, suck our blood, cause disease, destroy or foul our foods, attack our ornamental plants and lawns and sometimes simply cause fear, anxiety, and anger. This book will familiarize you, the homeowner, with the majority of household pests encountered in North America. It also provides you with the currently acceptable strategies and tactics to prevent their establishment in your home and how to safely and economically control them, once they have gained entrance and settled.

The book is organized into sections, each section dealing with a particular class of pests. For example, the first section deals with rather large pests all of which are warm-blooded, i.e. mammals and birds. Several of these pest species are simply pests on and around homes while others are either seasonal or permanent invaders of our living quarters. Some are mainly just nuisances that strew our garbage around, wake up the family dog, disrupt our sleep and foul our living spaces with their messy droppings and smelly nests. Some actually do damage, contaminate our foods, destroy our belongings and cause disease. Most of these wild, warm-blooded animals have pests of their own, fleas, mites, ticks and a variety of blood-sucking bugs. Unfortunately, their pests can also become our pests. Indeed the human population has experienced millions of deaths due to plagues transmitted by rodent-borne blood-sucking animals. Finally, these warm-blooded pests are annoying and frequently viewed with repugnance, fear and horror. Our loathing of rodents is in part cultural and our fear hardly well deserved. It certainly isn't rational to see a full grown adult human weighing 100,000 grams flee in terror from a 50-gram mouse. However, fear is frequently irrational and human nightmares are filled with common household pests, rats, bats, spiders and their kin.

The second group of pests includes those that bite, sting or suck blood. Some of the more dangerous of these animals, found in or around the home, sting or bite only when accidently encountered and their reactions are purely defensive and without malice. Nevertheless, their bites and stings are almost always quite painful and sometimes life-threatening or lethal. Others in this group are malicious hunters that seek out and prey on humans. These are the blood-sucking pests: fleas, mites, ticks, mosquitoes, bedbugs and their kin as well as some flies and midges. While the amount of blood taken by these small cold-blooded invertebrates is inconsequential, their aftermath—itchy and sometimes painful welts—torments us and in some cases cause bacterial, viral and parasitic infections.

The pests that invade our kitchens we find particularly distasteful, but our kitchens and pantries with their stored foods attract a variety of pests: ants, roaches, flies, grain moths, beetles, silverfish, etc. They don't eat much, but they do foul or damage foodstuffs, sometimes transmit disease and are generally annoying merely by their presence.

There are yet other pests that cause us no direct harm, but wreak havoc to our valued fabrics, furs, upholstery, rugs, books and papers. The economic losses they can cause might be considerable. These pests include a variety of carpet beetles, boring beetles, clothes moths and other animals that eat animal and plant materials. However, the greatest economic losses are caused by still another group of animals, those that can digest wood or bore into wood. The wood-eating and wood-boring pests literally cause more than a billion dollars worth of damage each year. The most notorious of them are the termites, but carpenter ants and wood-boring beetles also do their share of damage.

The last two categories of pests attack living plants, indoors and out. Today's use of decorative and sometimes expensive ornamental houseplants introduced the homeowner to a new myriad of pests (mealybugs, whiteflies, aphids, cutworms, spider mites, scale insects), that can and do kill ornamentals. They also frequent greenhouses along with slugs, thirps and snails. Finally, there is one group of pests that spares the apartment dweller and attacks only homeowners with lawns. The amount of effort needed to create a beautiful green lawn is enormous and the end result pleasing but all this can be undone by insects that feed on grasses:

beetles, chinch bugs, armyworms, sod webworms, etc. The only group of pests not included in this book are those pests that attack gardens and trees. However, a number of books dealing with garden pests are available.

In preparing this book my primary purpose was to provide the homeowner with a practical guide for pest control, but I had a secondary secret purpose in mind. As a biologist I find many of the so-called animal pests fascinating and even charming creatures. Their capabilities, natural history and rich history of interaction with humans make them, despite their depradations, an interesting topic for study. This being the case, I've attempted to incorporate some of the more intriguing biology of these animals in the text with the hope that the reader will spend some time watching these creatures before reaching for the can of Raid. Thus, the book is also a field guide to wildlife of the house and home. With it you can enjoy a convenient, interesting and inexpensive safari.

PART I
Warm-Blooded Pests

BATS

Bats, mammals of the order Chiroptera (meaning "hand-wing") have existed in their present form for at least 50 million years. There are approximately 850 different bat species known world-wide, and in total numbers they probably outnumber all other groups of mammals. Most bat species live in tropical regions, and only 39 species are known from the North American continent north of Mexico.

The news media (movies, television, etc.) often perpetuate myths, "old wives' tales," folklore, legends and fears about bats that a surprising number of people believe. Bats do not get into your hair. They are not flying mice. They don't come "out of hell." They are not blind. And only three species (none in the U.S. or Canada) subsist on a diet of blood. Myths about bats are found in many human cultures. The ancient Egyptians believed that bats could prevent or cure poor eyesight, toothache, fever and baldness, and a bat hung over the doorway of a house was thought to prevent the entry of demons that carried these "diseases." Bat gods were important in many pre-Columbian civilizations in Central America, and bats are used in voodoo worship in parts of Africa as well as in many parts of the Caribbean even today. The association of bats with the legend of human vampires has an uncertain origin, but since the time of Cortez and his conquistadors, peoples of Western civilization have linked vampire bats with the legendary "human" vampires of Transylvania. The writings of William Shakespeare, Robert Louis Stevenson and others have contributed to legends that cast a veil of fear over people through the association of bats with graveyards, death, ghosts and goblins.

To the Chinese, bats are regarded as symbols of happiness and good fortune (health, wealth, serenity, virtue and long life), and at one time Chinese mothers would sew small jade buttons in the shapes of bats on the caps of their babies, a custom believed to impart long life. Ancient and modern-day Chinese art objects, tapestries, Imperial robes and home furnishings often include bats as part of the motif.

1

In temperate North America bats feed almost exclusively on insects, and in the warm months of the year these ravenous nocturnal hunters may eat up to one-half of their body weight on a given night. If this level of consumption is extrapolated to a population of 50,000 bats (a conservative estimate for the number of bats living in a 100-square-mile area in New England), this would amount to over 13 tons of insects eaten in one summer. Included in this diet are many insect pests, including mosquitoes, biting midges, beetles and moths. While bats do on occasion carry rabies and may become minor pests when they roost in your home they are invaluable allies in the fight against flying insect pests. Some communities have actually constructed bat roosting sites to recruit more bats into their area.

Bats in the Belfry (and Attic)

Bats typically seek shelter in roosts during the daytime and are active on the wing at night, leaving their roosts shortly after sunset and returning before sunrise. The timing of their nightly comings and goings is closely synchronized with light levels and thus varies with seasonal changes in day length. One or more feeding periods may occur on a given night, depending on food availability and the temperature of the night air. Feeding activity is often interrupted by periods of night roosting, when individuals seek temporary shelter (alone or in small groups) in open buildings, or on the rafters of porches, carports and breezeways.

Many bats are gregarious animals who may seek daytime shelter in a variety of man-made structures, where they cling to walls, ceilings and rafters. When roosting they usually hang upside down with their wings folded next to the body. Bats that roost in small crevices, however, ordinarily assume a horizontal posture.

American bats usually give birth to a single annual litter of one or two offspring that may weigh from 20 to 30 percent of their mother's weight. This is comparable to a 100-pound human female giving birth to a 20- or 30-pound baby. Female bats suckle their young with milk (what else?) and weaning usually occurs at the age of from four to six weeks. Most bats grow rapidly and reach 90 percent of adult size by the time they are weaned. While they are not particularly prolific, they may live for several years, and because they have relatively few natural enemies, their numbers can become quite large.

SAVE THAT BAT: In Appreciation of Bats

One of the most difficult chores for an airborne predator is to intercept and successfully engage a moving target. To achieve this, an almost continuous ˉstream of information must be acquired, giving instantaneous data about target velocity, direction, and altitude. These data are then integrated with the attacker's speed and direction, and an intercept course determined. In the history of aerial combat, success is measured in kills, and to achieve the status of ace, five kills must be recorded. The greatest pursuit pilots of all air forces, the aces of aces, rang up some remarkable figures, perhaps the most—about 80—being recorded during daylight by Baron Manfred von Richthofen in World War I. This achievement is puny when compared with what an insectiverous bat achieves in its lifetime. A single bat can intercept and kill 900 insects in one hour and as many as 3,000 in one night's mission. Figuring a lifespan of fifteen years, in its lifetime a bat would have scored an amazing five million to ten million kills. Perhaps today's fighter aircraft with their heads-up display, guided missiles, skilled pilots and on-board computers have greater killing efficiency than the Red Baron, but they still can't match up with the night-hunting bats.

How do bats achieve this remarkable dogfight success? The bat begins its patrol at dusk, cruising at speeds of up to 40 miles per hour; it ranges the night skies with its sonar system emitting pulses at sound frequencies of up to 230,000 cycles per second, more than ten times higher than the human ear can detect. While cruising, pulses are produced at the rate of ten per second by the bat's voice box (larynx), which has large, ossified cartilages attached to strong muscles. Tension produced by this lever system applies tension to two small, vibrating membranes which produce the sound pulses, each of which lasts only a thousandth of a second. If the sound wave strikes an object, an echo is reflected back and picked up by the bat's extraordinary ears, which are structured to amplify and concentrate the sound.

Large echoes indicate a large target, small echoes a small target, and the crispness of the echo indicates target texture. If the echo returns at a lower pitch, the bat's doppler analyzer knows the target is moving away; if the echo is at a higher pitch, the target is approaching. Each ear picks up the echo at a slightly different intensity and time. Thus, the bat's minuscule on-board computer, its ten-milligram brain, automatically triangulates and makes in-flight directions for the intercept. As the bat closes on target, it increases its pulse rate up to 300 clicks per second and may even vary the tone (frequency modulation) to get more information echoes. When at point-blank range the target may attempt sharp evasive maneuvers but these changes are picked up and course corrections are made. The bat is so maneuverable that it can execute a 90° turn in a distance slightly greater than its body length. The bat then scoops the insect up with its wing or tail webs and stuffs it into its mouth. During this time, some bats that emit pulses through their mouths, stop sending hunting pulses, but others that can send pulses through their noses and still can acquire new targets while munching their dinners.

The entire hunt from first echo to kill takes only three to four seconds. Bat sonars' power-to-weight ratio makes it about a billion times more efficient than our best radar or sonar detector systems. It can even pick its target out of a number of decoys. A. Donald Griffin, who discovered bat echolocation, showed that with a little training bats could pick out of the air a mealworm that had been thrown up with a handful of pebbles. How our antiballistic missile engineers would like to match that feat in picking up decoy warheads from the real ones! How efficient is the bat as an interceptor? Even moths that can detect bat sonar and take evasive action are hit 60 percent of the time, and insects without sonar detectors may fall victim 90 percent of the time.

While some species of bats migrate to warmer climates when summer is over, most American bats winter over, seeking shelter in homes, lofts, and other places protected from freezing temperatures. During this time they go into hibernation and their body temperatures drop, their heart rates decrease from 1,300 beats per minute while in flight to 20 beats per minute. During this hibernation period, bats survive by burning up the fat reserves that were accumulated in late summer and early fall. However, hibernating bats may rouse periodically from the hibernating state during unseasonably warm winter periods or when repeatedly disturbed by humans.

TYPES OF BATS THAT FREQUENT HOMES

The Little Brown Bat (*Myotis lucifugus*)

The little brown bat is one of the most abundant of all colonial bats in the northern part of its range, which runs from southern Alaska to all but the southernmost states. In the spring and summer, females form maternity colonies of hundreds of individuals or more in attics, barns, and other retreats that are dark and hot during the daytime. In winter, these bats hibernate in caves and mines, frequently returning year after year to the same nursery

Little Brown Bat

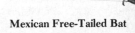

Mexican Free-Tailed Bat

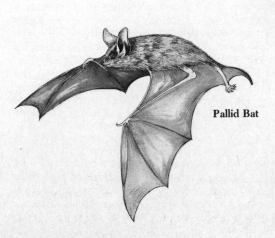

Pallid Bat

colony and hibernation cave. Colonies are far more common near lakes and rivers. Rabies is seldom a problem with this species, since it is not an effective biter due to its small teeth.

Little brown bats are covered with a dense, fine, glossy fur. Both sexes are rich brown—almost bronze—in color, and juveniles may be almost black. Their ears and membranes are a glossy dark brown. There are many small brown species, but this is the one most often found in buildings. They weigh about 1.4 ounces and have wingspreads of up to ten inches.

The Big Brown Bat

The big brown bat, *Eptesicus fugus,* also called the house bat or barn bat, is probably the largest bat commonly found in buildings except for the pallid bat. Most adults are copper-colored, but color may vary from light to dark brown. Each hair is bicolored—the basal half being almost black and the outer half brown. Its face, ears and membranes are dark brown to nearly black. It weighs up to 8 ounces and has a wingspread of up to 16 inches.

The big brown bat is probably the colonial bat most familiar to man. In summer it commonly roosts in attics, belfries, and barns or behind awnings, doors and shutters, but seldom in caves. It is a hardy species that can endure subfreezing temperatures but is not as tolerant of high temperatures as is the little brown bat. During hot weather, it may crawl into rooms from crevices of fireplaces, or both young and adults may appear in basements if the space between the inside and outside walls is continuous from attic to basement. Colonies vary in size from 12 to 200. These bats have a remarkable homing instinct but do not migrate far from their place of birth. They are one of the last bats to hibernate in fall and the first to rouse in spring, and may be seen flying about at dusk in late November and early March, spending the winter in buildings, caves, mines, and similar shelters. *Eptesicus* is easily recognizable due to its large size and steady, straight flight at a height of from 20 to 30 feet or more. After feeding, the bat flies to a night roost to rest, favoring porches, brick houses, garages with open doors, or breezeways. The tell-tale signs of its presence are a few droppings left each morning below the roost. Big brown bats can inflict a painful bite if carelessly handled. This is one of the species that is most often rabid.

The Mexican Free-Tailed Bat

The Mexican free-tailed bat, *Tadarida brasiliensis,* is a rather small bat with long, narrow wings, and about one-half of the tail extending beyond the interfermoral membrane. The ears almost meet at the midline but are not joined, and have a series of papillae, or wartlike structures, on the anterior rims. The upper lips are wrinkled. The body and membranes are dark brown.

They are found from California to Florida, migrating into Texas and Mexico, and occasionally are found as far north as Oregon, Nevada, Utah, and, in the East, as far north as North Carolina.

This species is the most colonial of all bats. The habitat of the free-tailed bat differs in various parts of the United States. It inhabits buildings on the West Coast and in the Southeast. It is primarily a cave bat of Arizona, New Mexico, Oklahoma and Texas. Maternity colonies of 1,000 or more may inhabit a single building in California. In Florida the species never enters caves, and thousands have been found in a single building.

The Pallid Bat

The pallid bat, *Antrozous pallidus,* is light yellow above, the hairs being tipped with brown or gray, whereas the underparts are a pale cream, almost white, and its membranes are tan. This is a large bat with big ears, large eyes, broad wings and distinctive, pig-like snout. They are found primarily in the western United States from the Pacific Northwest to the Southwest.

This colonial species is occasionally troublesome in California, where the same open shelter serves as both day and night roost. The bats hang from the rafters and their droppings foul hay in barns and cars in garages. This species has one of the most unusual feeding habits of any North American bat: prey is mainly from the ground and little, if any, food is captured in flight. Food consists of scorpions, grasshoppers, Jerusalem crickets, June beetles and other ground beetles. It is a relatively slow flier, foraging close—0.9 to 1.2 m (three to four feet)—from the ground. It has been found rabid, but is rarely a health problem to humans.

HOMEOWNER'S BAT PROBLEMS

Bats may attempt to bite if handled—and they should not be handled unless you are wearing heavy gloves, since rabies can be transmitted to humans by bites. In temperate regions of North America, the incidence of rabies in most colonial bat species is less than one-half of one percent, but the incidence is somewhat higher (4 to 10 percent) in sick bats. If bitten by a bat, an effort should be made to capture or kill it and have it tested for rabies. A wound from a bite should be washed thoroughly with soap and water, and *you should immediately contact a physician* and the public health department. Antirabies treatments are usually mandatory. (see Rabies, page 27).

Bats may also transmit histoplasmosis, a fungal disease of the lungs that grows in moist bat droppings (guano). The pneumonia-like symptoms are caused by the inhaled histoplasmosis spores. In many parts of the United States, this fungus occurs naturally in the soil, and residents of these areas develop a natural immunity. However, the prolonged presence of bats in an enclosed area or building may increase the risk of histoplasmosis in certain parts of central and southeastern United States.

While roosting in buildings, bats may emit audible clicking sounds that can be annoying to some people. Sometimes they may be heard clamoring in the spaces between chimneys and walls or between walls. Bats do not gnaw or chew wood, metal, or plastic to gain entry or exit from buildings. Their roosting places in buildings are often recognized by the presence of a whitish or darkened stain on rafters and the accumulation of fecal material (guano) beneath the roost site. Because bats also defecate and urinate while in flight, fecal droppings and drops of urine may become splattered on the outer wall of a building near where the bats gain entry. A musty odor is often associated with the presence of a colony of bats in a building, often due to the accumulation of guano and to the odors of the bats themselves.

CONTROLLING BATS IN THE HOME

Bats sometimes take up residence in, or accidentally enter, houses and other man-made structures. To evict a colony of bats from a house, the only environmentally sound, effective and permanent method is to close the openings that bats use for their exit and entry. Bats usually enter buildings through small openings along the edge of the roof, through crevices where masonry has pulled away from the clapboard, or through unscreened air vents. To determine where bats are entering or exiting a building, two or more people should position themselves outside the house shortly after sunset (in the warm months). Bats can best be observed as silhouettes against the twilight sky. It may take up to one hour for all bats to depart on a given evening. Usually they will not fly out on cold, rainy nights. If you see something flying in and out of your chimney, you probably don't have bats; you most likely have swallows or chimney swifts instead.

After you have located places that bats are using as exit and entry holes—and all the bats have departed to feed at night—the openings should be sealed so that the bats cannot reenter. It may be necessary to repeat this procedure several times until all openings have been closed. The best time to seal the openings is in the early spring or in fall. One should avoid sealing openings from late May to mid-July, since bats too young to fly may be trapped inside. Chemical repellents (e.g., mothballs, sulfur candles, ammonia) are generally ineffective in getting rid of bats. Bats may be temporarily evicted by use of these repellents, but they are likely to return after the chemical has dissipated. Pesticides (e.g., Rozal, chlordane, lindane, etc.) often sicken bats before killing them, and they are seldom completely effective. When these poisons are used, bats can become sick and fall to the ground, increasing the risk of exposing children and pets to sick bats. Application of poisons to the interior of a house for controlling bats may impose health risks to humans; consequently, this practice should be avoided. Devices that produce sounds at ultrasonic frequencies (sonar type) are not very effective in ridding buildings of bats. Use of lights in roosting areas may reduce the number of bats under some circumstances. The safest and soundest approach for eliminating bats from buildings is still to physically close the openings that bats are using as entrances and exits. Of course, this may be difficult to accomplish if the building has numerous holes and crevices that bats can use.

If a single bat gets into a house accidentally (a common occurrence in late summer, when young bats are learning to fly), the lights should be turned on and doors or windows should be opened. The bat will soon leave on its own. Unless the bat is sick or injured, it will invariably leave, not to be seen again. If a bat is discovered in living quarters during the day, efforts to encourage it to leave will be most successfully undertaken if left until nightfall.

HOUSE MICE

The house mouse, *Mus musculus,* has been a common pest to man for thousands of years. *Mus* is a derivative of the ancient Sanskrit word "musha," meaning to steal, and *musculus* reminds us of muscular—a highly appropriate characterization of these wiry little thieves. The term "commensal" is also used to refer to the house mouse; it means "sharing the table," which also seems appropriate, although they are certainly rarely invited dinner guests.

The house mouse, despite its pilfering ways, held religious significance in the Orient. In Japan it was considered the messenger of the god of wealth. Throughout the Orient, the mouse has found recognition in that one of every twelve zodiacal years is the Year of the Mouse (or Rat).

Mice have also found their way into history when they helped the Allied effort in World War II. During the siege of Stalingrad, German tanks were stored in pits and camouflaged with straw

House Mouse

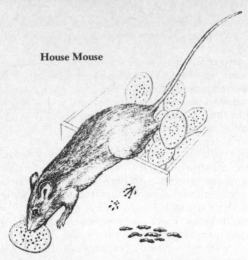

while waiting in reserve. When a Russian counterattack seemed imminent, orders were given to mobilize, but most of these tanks couldn't be started—mice had eaten away all the wire insulation and shorted out the tanks' electrical systems.

In North America, house mice are found throughout the United States, southwest Canada north to central British Columbia, along the coast of Alaska, and throughout Mexico. This ubiquity is due to the amazing fertility and adaptability of the Old World rodents (including the house mouse). House mice were introduced into the New World in the early sixteenth century by the early explorers and colonists. The house mouse usually has a grayish-brown body that is a bit lighter on the underside. It has an average adult length from tip to tip of from about 5⅛ to 7¾ inches, and the tail length is anywhere from 2½ to four inches. There are no coloration differences between the sexes, and they are about the same adult weight (⅝ to ¾ ounce).

Another distinguishing feature of the house mouse is its ungrooved incisors. These incisors are constantly growing, at the rate of four to five inches per year. Despite statements in many books, mice do not need to gnaw continuously on objects to grind down their teeth.

House mice reach sexual maturity between 1½ to two months of age. The reason for their fecundity is simple: although most broods are produced in early spring and in late summer, mice can breed at any time of year and they often bear five to eight litters (four to seven pups per litter) in a year. Such prolific reproduction is possible due to a short gestation period—only nineteen days.

Mice are active throughout the year at any hour of any day, but seldom stray far from their nests, having an average home range in your house of only twelve feet. These limited home ranges are marked by urine and secretions, and new objects are marked in the process of their explorations. This is why traps should never be thoroughly cleaned after use, because that removes all familiar odors.

Unlike rats, house mice are nibblers or snackers, feeding up to twenty times a day, and preferring sweet and high-protein foods, particularly cereal grains, but will also munch on glue, soap, leather, meat, plastic and wood. They do exhibit preferences and, once they have a favorite food, they stay with it and are slow to change their tastes. For this reason, pre-baiting is suggested to increase the success of poisoned bait.

Mice, unlike rats, are able to get along without water, since they get their water from moist food and from metabolism. They will, however, drink readily if water is available, and sweetened water baits are good delivery systems for rodenticides in dry climates.

A mouse must consume three to four grams of food per day (about 10 percent of its body weight), but may carry away as much as it eats, since it often hoards food in its nest. Mice, being nibblers, waste a good deal of food; and what they don't eat they often contaminate with their droppings, urine and hair. As mice dart around picking up morsels, they leave their calling cards—cylindrical, brown, pointy-ended fecal pellets. In the course of a year, one house mouse can drop as many as 18,000 of these pellets—and that's no mean accomplishment. As a matter of fact, the best way to tell if you have a mouse problem is to sweep up the droppings. If new ones are there the next day, you do have a problem mouse.

Before discussing problems that mice cause, a few bits of interesting mice trivia. Mice can jump up to 12 inches. I've thrown a mouse out of a sixth-floor window and it walked away. They are great climbers. They swim; and even flushing one down the toilet is no guarantee that it won't re-surface in a minute or so. They can squeeze through holes about a quarter of an inch in diameter. They have an average life span of about one year, and they have very good memories.

There are many mouse-associated diseases, including: plague, murine typhus, rickettsial pox, trichinosis, salmonellosis, anthrax, glanders, tularemia, Rocky Mountain spotted fever and rabies.

While this list is impressive, the house mouse is seldom a problem—except for food poisoning—but wild mice of other types that may get into your home are the real problem animals. The main problem with mice is that they destroy and contaminate food, they gnaw holes in clothing and woodwork and sometimes cause fires by chomping on electrical wire insulation.

Prevention and Control: Mouse-proofing and good housekeeping together make a starting point for preventing and controlling mice. Mice can be excluded from the home by blocking possible entryways. Seal off all cracks or holes in floors, walls and foundations that are larger than a quarter of an inch in diameter. Larger openings can be closed with galvanized metal sheeting, hardware cloth or galvanized mesh wire, while smaller holes can be closed with cement or a good caulking compound. Once mice are established in the home, there are three strategies available to the homeowner that will eliminate them. First, the new commercially available glue boards are quite effective. Set up a covered bait station, e.g., a paper bag or cardboard tube, and tape the glue board inside it; this keeps the boards from getting dusty and losing their tackiness. Baits can be raisins, peanut butter, nuts or even bits of bacon. The second (and probably best) method is the use of snap-back traps. These are available at hardware stores and can be used over and over again. Use multiple traps, tie the bait to the trigger and place traps where there is evidence of mice (gnawing or droppings). Make sure to check the traps every day to remove the dead mice before they decay. Wear rubber gloves or use tongs when doing this and flush the carcasses down the toilet. Always wash the gloves or tongs thoroughly after such usage. Finally, there are commercially available poison baits.

A number of restricted, highly toxic baits are available to licensed pest control operators but not to homeowners. However, baits containing coumarin-like compounds are widely available. These compounds act by inhibiting absorption of vitamin K (which is needed by the clotting mechanism); the end result is that the mice will eventually die of internal bleeding. There are two problems posed by poison baits: (1) children and pets may get at them and (2) the mouse may expire in its nest in your walls. Dead mice and rats give off a nauseating odor as they putrefy that may persist for day or even weeks. The second problem with dead mice in your walls is that the carcasses attract other pests such as cluster flies and dermestid beetles, which will invade your home as soon as they have finished dining on the dead mouse.

MEADOW MICE AND THEIR KIN

Meadow mice, field mice and deer mice are occasional visitors of homes and gardens. There are several species that can be pests. The common meadow mouse, or meadow vole, *Microtus pennsylvanicus,* is common throughout Canada and the United States wherever there is good ground cover. These mice construct a maze of narrow (one- to two-inch-wide) runways in matted grass, and in the process may chew up parts of your lawn. They are from three and one-half to five inches long, have short, one-and-one-quarter- to two-and-one-half-inch-long tails and weigh one to two and one-half ounces. Their thick, long, soft fur ranges in color from gray to brown above and silver to gray below. Their reproductive capabilities are extraordinary, producing as they do as many as nine litters of five or six pups in a single breeding season (March to November). In summer, meadow voles can damage growing plants and during the winter may do damage to trees and shrubs.

The pine mouse, *Microtus pinetorum,* is a small-bodied cousin to the meadow vole, being two and one-half to four inches long, with a short (two-thirds- to one-inch-long) tail. Its eyes, unlike those of the meadow vole, are sunken. It's a handsome little critter, covered with soft, thick, auburn-colored fur. It is abundant in the eastern half of the continent, usually associated with forest floors. These pine voles are active day and night and, unlike the

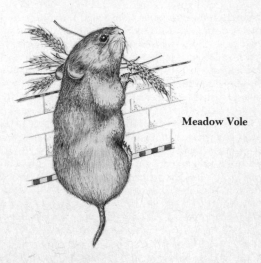

Meadow Vole

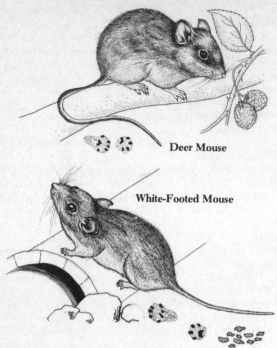

Deer Mouse

White-Footed Mouse

surface-dwelling meadow voles, construct tunnels and, in the process, damage roots and eat valuable bulbs.

Deer mice, *Peromyscus maniculatus,* are attractive little mice (two and one-half to four inches long, with two- to five-inch-long tails). These mice are widely distributed and sometimes, to escape the cold during winter months, take up residence in homes, where they damage foodstuffs and tear up furniture to get nesting materials. They range in color from gray to red-brown and have bicolored tails. They are very similar in appearance to another widely distributed pest, the whitefooted mouse, *Peromyscus leucopus.* Both of these species are prolific and build burrows in the ground and in buildings, where they become real pests.

Control: See mouse control tactics on page 12.

RATS

There are two species of rats that plague homeowners: the Norway rat and the black rat. The Norway rat, *Rattus norvegicus,* also called the "barn rat," "sewer rat" and "wharf rat," is a large, robust, cunning, nocturnal mammal that was introduced into North America by the early colonists. It has thrived and spread to most urban and suburban parts of the United States and Canada, where it has become a major pest species, costing millions of dollars in damage and other millions in control efforts.

An adult rat averages about sixteen ounces, but some get to be twice that weight. This rat has a blunt snout, small eyes, small ears covered with dense, short hairs. Its upper coat is soft gray-brown with scattered black hairs, while the fur on the belly is shaggy and gray to yellow-white in color. Its tail, which is shorter than the head and body length, is dark above and pale beneath. As it moves about, it leaves its calling cards: blunted, capsule-shaped, dark brown, 4/5-inch-long fecal pellets. It also announces its presence by: distinctive footprints and tail marks (see pages 15 and 16); earth excavated from its burrows; dirty, greasy marks and discolorations left along its runways; "gnaw" marks made by its sharp incisor teeth and sometimes a telltale foul odor. These rats are agile climbers and excellent swimmers. Unlike house mice, rats must drink and hence are often found near water.

They usually construct their nests outdoors, under slabs, garages, rock piles, next to foundations, etc. The nests are underground burrows composed of a central nesting area and several tunnels. The main entrance is open, but several alternative escape tunnels have their exits loosely covered so that the rat can bolt when its tunnels are disturbed. These rats are long-lived (from three to four years) and prolific breeders, producing four to seven litters of eight to 12 pups each in a year. Gestation is about 21 days; the pups are weaned and on the prowl by the time they are forty days old. While their range varies, it has been found that city-dwelling rats usually stay within 60 feet of their nests. Norway rats also can set up housekeeping in *your* home, often getting between walls and in crawl spaces. These animals are omnivorous

Norway Rat

Black Rat

and will exploit a great variety of foods if they are available, including meat.

The black rat, *Rattus rattus,* also called the "roof rat" or "Alexandrine rat," is a sleek, dusky black, extremely quick, agile night prowler that arrived here by ship in the 1600's and still tends to be more numerous in coastal areas. It has a sharp, pointed snout, weighs only eight to 12 ounces, has very large, stiff, almost naked ears, large, prominent eyes and a dark tail longer than the length of its head and body. The roof rat's fur is stiff, gray to black above and white, gray or black below. Its droppings are about half an inch long and pointed at the ends. The roof rat is a marvelous jumper and excellent climber that frequently gains access to homes through the eaves, reaching them by climbing over tree branches, vines or wires. In the home it builds its nests in the walls, attics and crawl spaces, where it produces four to six litters of up to six pups each in the course of one year. This black pest is restricted to the southern and southwestern United States.

Historically, the black rat is one of the greatest animal villains of all time, for it was the main vector of the plague, "The Black Death," that swept over Europe in the fourteenth century, causing at least 25 million human deaths. It was also responsible for the plague outbreak in Asia early in this century that claimed another eleven million human lives.

Rats, with their great natural cunning, remarkable adaptability and high reproductive rate, are one of the most unrelenting and destructive animal pests. They pollute and destroy foods, gnaw plastic, paper, books and insulating materials; they spread disease

The Black Plague

One-third of the population of Europe was killed by the Black Death during the Middle Ages. The cause of this devastating epidemic was a bacterium that used to be called *Pasturella pestis* but is now called *Yersina pestis* after its discoverer, Dr. Yersin. This microbe infects a number of rodents; the animals' bloodsucking fleas pick up the infection and can in turn transmit it to humans. In humans, the disease can manifest itself in several forms: (1) Bubonic plague—where *Y. pestis* invades the lymph nodes of the armpits, groin and neck and produces swollen, pus-filled, black, hemorrhagic lumps (buboes) as large as eggs. The mortality rate from this form of the disease is up to 70 percent. (2) Septicemic plague is a more deadly form of the same disease and causes many small subcutaneous hemorrhages that turn the skin black; hence the name "Black Death." (3) Another manifestation of this microbe is pneumonic plague. This affects the lungs and is rapidly spread via aerosolized droplets that are coughed up. In this form it can spread from human to human and is almost always fatal. Finally, there is sylvatic plague, which is now endemic in rodents (ground squirrels, wood rats, deer mice, marmots) in the western United States. It is not very contagious, since we have little contact with these animals, but it does represent a reservoir for *Yersina pestis*.

and reduce real estate values; they cause fires by shorting out electrical wires and cause asphyxiation by biting through gas pipes. One disease, rat bite fever, is a frequent aftermath of rat bite. Rat bite is the fourth most common bite reported to emergency rooms, exceeded only by dog, cat and human bites. Unfortunately, the usual victims of rat bite are infants. Among the other diseases carried by rats and transmitted to humans bitten, directly or via the rats' ectoparasites, are murine typhus and food poisoning.

Prevention and Control: The presence of rat droppings, rat tracks and marks of gnawing are excellent indicators that you have a problem. First, look around the outside of the house, since the Norway rat most often constructs its nest outside, in underground burrows near buildings. These burrows typically have two entrances and several lightly covered or hidden emergency escape exits. Finding an entrance, particularly if there is fresh earth or track marks, indicates that you have several rats present that may gain entrance to your home. Start by denying rats entrance to the house. Close openings around pipes with galvanized metal sheeting or concrete; cover ventilator grills with wire. Wooden doors and windows to the outside can be gnawed by rats and should be protected with galvanized metal sheeting (24-gauge). Be particularly careful to close openings to hollow walls and higher eaves to attics.

Deny rats easy food access by keeping all garbage containers (indoors and outdoors) tightly covered. Indoors, packaged foods should be kept in tightly sealed cabinets and individual metal containers. Make sure to sweep up and vacuum food remnants after meals and make sure that young children don't leave tempting morsels about. The family pet's food dish is also a potential source of rodent food, since some pets are messy eaters while others are very picky and leave food uneaten. This is an open invitation to rats and mice. Blocking access and denying food is the best way to start, but if you continue to find tracks and droppings, a more active control program is called for.

Active Rat Control by Commercially Available Methods: The best strategy for the homeowner is to use the old standby,the snap-trap. Trigger snap-traps are readily available and inexpensive. Use a lot of them—a dozen or more. Tie the bait to the trap, place baits along runways where there are tracks, markings or droppings, or where there is evidence of gnawing. Baits can include peanut butter, dried or smoked meats, nuts, even cheese. Do not use snap-traps if there are small children or pets about.

Some of the recent technological advances, such as glue boards and ultrasonic sound generators, are not particularly effective in controlling rats; the only other choice is poisoned baits. And the only poisoned bait that should be used by homeowners is an anticoagulant-type bait. These were discussed on page 12, under mouse control. While these are effective, the aftermath—dead rats in your walls—is not pleasant (see page 15).

If you locate burrows next to your foundation, they can be fumigated with carbon monoxide by means of a hose run from your car's exhaust. Push the hose into the opening and seal with soil or sod. Locate hidden escape holes by the smoke that will arise from them. Seal these too. Make sure there are no holes in your basement wall that will allow the toxic fumes to get into the house.

If you have a major rat problem, call for professional help.

CHIPMUNKS

Chipmunks are gnawing mammals belonging to the order Rodentia. Like all rodents, they have a pair of opposing frontal incisor

Chipmunk

teeth for gnawing, and grinding teeth along their cheeks. They belong to the squirrel family, Sciuridae; thus they are closely related to prairie dogs, woodchucks, ground squirrels and tree squirrels. There are a number of species of chipmunks widely distributed throughout the United States and Canada.

Chipmunks are small (three-ounce, nine-inch-long) squirrel-like animals that scurry about during daylight hours with their three- to four-inch-long tails held upright. Some look as if they have mumps, as their cheek pouches bulge with nuts being carried to their hoarding areas. Often, chipmunks announce their presence with a sharp chuck-chuck-chuck noise. Eastern varieties have three black stripes running down the back and sides to the rump, while western species have five black stripes. All chipmunks also have white facial stripes. Their fur, while a bit variable in color, is most often reddish-brown.

Chipmunks live in burrows and will sometimes build under homes or garages or may gain entry to homes at ground level. On occasion they may raid strawberry patches or dig up your expensive bulbs.

In the western United States, chipmunks have been found to carry plague bacteria, a highly virulent disease that can be transmitted by infected rodents or the fleas that inhabit their fur (see page 16). "Chippies" have also been shown to harbor a number of other disease organisms that can affect humans, either by direct contact or via the family cat—if it is a good hunter. The risk these friendly little animals pose is relatively minor and they are pleasant neighbors unless present in large numbers.

Control: Whenever chipmunks become a nuisance in dwellings, use the same procedures described for controlling other small rodents (see page 12 and above). This includes use of repellents, live traps, spring traps, burrow fumigation and poisoned baits.

TREE SQUIRRELS

There are a number of tree-dwelling squirrels found throughout the United States and Canada. They are probably the most familiar small mammal, since they are highly visible daylight foragers and can thrive even in overbuilt urban areas as long as there are trees present. They tame easily, are friendly, noisy, bushy-tailed, active, pleasant little animals that in the spring engage in noisy, acrobatic territorial squabbles. In some parts of the country they are actively hunted as game animals; in many states there are specific squirrel-hunting seasons. They are marvelous climbers and leapers and, like the chipmunks, often hoard more food than they need.

They usually nest in trees, but unfortunately sometimes find humans' homes equally acceptable. They get in under eaves or through air vents or broken windows and store food and build nests between walls and in attics. They are even known to build nests in chimneys, which results in a smoky mess when you build your first fire of the winter. With their gnawing teeth, they often do considerable damage to wooden shingles and mouldings of windows. Squirrels also have the habit of gnawing on electrical wires and telephone lines, causing outages and, in some cases, fires due to the resulting short-circuit. Why they gnaw electrical cable is not known for sure, but it may be to get nesting material.

Red Squirrel

Gray Squirrel

If in your overhead or in your walls, squirrels are frequently noisy, disturbing you and—worse yet—the pet dog, who begins to bark. Because of their friendliness, people often get close to squirrels, but they are unpredictable wild animals that will on occasion be quite aggressive and bite, seemingly without cause.

Among the more common squirrels are the red squirrel, *Tamiasciurus hudsonicus*, the eastern gray squirrel, *Sciurus carolinensis*, the western gray, *Sciurus griseus*, and the southern and northern flying squirrels of the genus *Glaucomys*, whose nocturnal glides often end with a noisy landing thump on your roof that will awaken even the soundest sleeper. Mature gray squirrels have a head and body length of from eight to 12 inches and bushy tails of almost an equal length. They weigh between three-quarters of a pound to a pound and a half. Usually they are grayish with

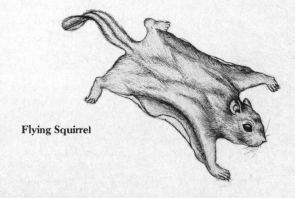

Flying Squirrel

lighter undersides, but darker and lighter variants exist. Flying squirrels are small, having heads and bodies five to six inches long and three- to four-inch-long tails. They have thick, glossy, olive-brown fur (dark above and light below) and a folded layer of loose skin along the sides of the body that is attached to the front and rear legs. This wing-like structure is stretched out to provide an air-supporting surface used to make long glides.

GROUND SQUIRRELS

There are several species of ground squirrels of the genus *Spermophilus* that are occasional pests of houses west of the Mississippi. They are long-lived, ground-burrowing, gregarious animals that sometimes form large colonies. Adults may range in size from one to two pounds. They sometimes burrow under buildings or

Ground Squirrel

may temporarily get into houses when foraging for food. They also can cause considerable damage to vegetable gardens. In the West they carry a number of disease organisms, including plague, tularemia and Rocky Mountain spotted fever; hence, dead ground squirrels should not be handled without gloves.

WOODCHUCKS, MARMOTS

Woodchucks, or "groundhogs" as they are commonly called, do not chuck wood but do chuck dirt—a lot of dirt—as they dig their rather extensive burrows. They belong to the family Sciuridae; thus they are cousins to the squirrels and chipmunks. There are several members of the genus *Marmota* found in the United States and Canada, but only one, *Marmota monax*, is called "the woodchuck." The rest are simply called "marmots." These animals have long been considered rural pests of orchards, where they strip

Woodchuck

Marmot

bark off sapling fruit trees. They also burrow in pastures, where cattle can break legs stepping into their holes. Woodchucks also like to invade vegetable gardens and, having ravenous appetites, can ruin a vegetable garden in short order. They *occasionally* are found in the suburbs of large cities, where they dig their burrows under houses, barns and garages. In the process of burrowing, they excavate a considerable amount of soil, creating an unsightly mess.

Woodchucks are heavy-bodied, squat, furry animals covered with light brown to dark brown hairs that often are frosted at the tips. The fur on the underside is paler, and sometimes there is a touch of white at the snout. Their bodies are 16 to 20 inches long; their stubby, fur-covered tails are four to seven inches long and they weigh between five and ten pounds. When frightened, they scoot for their burrows on relatively small brown-black feet. They emerge after hibernating all winter and start to eat at a furious pace to increase their body fat before the next hibernation. They are particularly active in the morning and often forage in late afternoon too.

Prevention and Control: Woodchucks in burrows under buildings can be fumigated out only if you are sure the poisonous fumes won't get into the house. A hose run from the exhaust of your car can be used to pump carbon monoxide into the burrow. There are also commercially available woodchuck cartridges that can be put into burrows and when ignited produce noxious fumes. To increase efficiency, locate and plug all exits from the burrow. Woodchucks are easily captured in baited humane traps (Havahart or Tomahawk). The use of baits poisoned with strychnine will get rid of woodchucks, but this is a dangerous, highly toxic substance, should be used with caution and in some areas can be used only by a licensed operator. In areas where the law permits, shooting with a .22 is quite effective, although, from my own experience, woodchucks quickly learn to head for cover once they hear a shot.

MUSKRATS

The only people who have trouble with muskrats, *Ondatra zibethicus,* are people with streams or ponds in their yards. These large aquatic voles make burrows in the banks of ponds and streams and, in the process, can cause damage and erosion. They feed on aquatic vegetation and can be destructive in lily ponds.

These two- to four-pound, large, furry brown rodents have bodies up to 14 inches long and naked, black, flattened (from side to side) tails up to 11 inches long. They once were hunted extensively as a fur animal. They can wander a long way from their dens, which are made of marsh vegetation and may be up to three

Muskrat

feet high. They are widely distributed throughout much of the United States, Alaska and Canada.

Control: The best way to control muskrats is to mow around the ponds and streams, cleaning up messy areas and removing aquatic vegetation which the muskrats use for food and shelter. If this doesn't get rid of them, get a license to trap them, using humane, cage-type traps that are commercially available. Bait the traps with pieces of apple or carrot and place them near burrow entrances.

RABBITS, COTTONTAILS

Widely distributed across much of the United States are several species of cottontail rabbits that are occasional pests of suburban yards. They damage our summer gardens by munching on succulent vegetables and in the winter may injure ornamental shrubs, fruit trees and berry bushes. Rabbits and their larger, longer-legged cousins the hares, belong to the family Leporidae and have in common long ears, long hind legs, soft fur, a lighter colored underside and short, cotton-like tail. The eastern cottontail, *Sylvilagus floridanus*, is typical. Its head and body length is up to 18 inches, its ears are three inches long and its weight is from two to four pounds. Cottontails are common wherever there is nearby heavy brush or semi-wooded areas. Their nests are shallow depressions lined with the mother's fur, grass and leaves. Litters of from three to six young are born from spring to fall, and within three weeks these are ready to leave the nest. Wild rabbits are generally harmless but do act as a reservoir for tularemia.

Eastern Cottontail

Prevention and Control: The best way to stop rabbit damage to gardens and ornamental shrubs is to use mechanical barriers such as wire fencing. A one-inch mesh at least 30-inch-high chicken wire fence held in position by stakes will provide protection for five or more years. And the cost is relatively small. Make sure the bottom of the fence is in the soil, to prevent rabbits from digging under it. Shrubs can be protected by cylindrical chicken wire guards or by painting them with a rabbit repellent. The repellent, powdered rosin dissolved in denatured alcohol, should be applied in the fall. Make sure to paint two feet higher than the snow will drift. Shooting is permitted in some rural areas; if there are only a few rabbits you can use commercially available box traps baited with apples or carrots.

MOLES

Moles are subterranean, small, furry mammals that spend most of their existence underground. They are occasional lawn pests because of the ridges and mounds they create as they push the earth up during their incessant tunneling. Their burrowing tools are their enormous, clawed, spade-like front feet. They have naked, pointy noses or noses with a variety of finger-like projections. They are earless and have tiny, pinhead-sized, almost sightless eyes. Their coats are extremely soft and thick. The most common of nuisance moles is the eastern mole, *Scalopus aquaticus,* which is up to eight inches long from snout to tail tip and weighs up to 120 grams. Moles feed on grubs and earthworms, some of which they store to carry them through the winter. They are fairly widely distributed but are not found in the Rocky Mountain or Great Basin areas.

Control: The best way to get rid of moles is to deprive them of their food supplies—grubs and worms—by treating the afflicted lawns with either milky disease spores, which kill Japanese beetle grubs, or diazinon granules (see page 169 on controlling lawn pests). There are also several different types of mole traps available

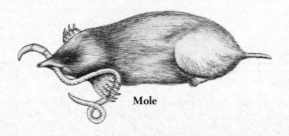

Mole

commercially, e.g., harpoon type, gopher trap type and choker loop type. These traps come with instructions for their use. They are effective but expensive.

Rabies

Symptoms: The symptoms of rabies infection have been in the literature since 1800 B.C., and few people are known to have recovered from this disease once clinical symptoms have appeared. The causative agent associated with the saliva is a bullet-shaped virus which enters through the wound, penetrates nerve endings, migrates within the nerve cells to the spinal cord, and from there to the brain. During this whole time, which may take weeks to months, the virus multiplies. From the brain, the virus moves out in all the nerves including those to the salivary glands, where the virus continues to multiply. The early symptoms include: a change in disposition, restlessness, and anxiety; this is followed by a stage of fury and aggressiveness. Dogs and other animals in this stage lose fear and bite anything nearby. As the disease progresses, the lower jaw droops, there is extensive frothy salivation, and there develops an inability to drink, possibly because it is so painful to swallow. Thus, hydrophobia, fear of water, is a common symptom. In the final stages of the disease, motor paralysis begins and movement becomes uncoordinated, as in severe alcoholic states. The cause of death is always respiratory failure due to paralysis of the muscles involved in breathing.

First Aid: None! If bitten by any wild warm-blooded animal (squirrels, skunks, foxes, bats) or domesticated animal (cats or dogs), see your physician immediately. Successful treatment must begin as soon as possible.

Medical Treatment: Until recently treatment consisted of 14 to 21 painful injections. Every year over a million people have been treated, and it is estimated that at least 30,000 people are treated annually in the United States. A new treatment involving only four to six innoculations has been perfected, and with this new vaccine there are relatively few side effects. Once again the treatment MUST be started within a short time of being bitten.

Tetanus

Tetanus can also occur from improperly cleaned, closed wounds. It is caused by the anaerobic bacterium *Clostridium tetani,* which secretes tetanus toxin. The toxin mainly affects the nervous system. The incubation period ranges from four to 21 days. Symptoms include headache, stiffness of the jaw (lock-jaw), muscle spasms, and rapid pulse. Death is usually a result of respiratory arrest. This disease can be prevented by proper surgical therapy of wounds and passive or active immunization. However, if the clinical tetanus appears hyperimmune, human tetanus anti-toxin, muscle relaxants, and antibiotics should be taken.

PORCUPINES

Looking like large black pincushions, porcupines, with their coats of long, sharp, barbed quills, are the largest rodents found in the United States and Canada. Their defensive armament is awesome, and although they are slow, clumsy and lumbering, they are relatively immune to attack. The porcupine's entire upper body and the upper surface of its powerful, muscular tail is covered with over 20,000 needle-like quills, some of which are four inches long. The quills are attached to muscles, and the porcupines, when threatened, can raise them at will. Contrary to popular belief, they can't shoot the quills at you, but they can twirl and slap with their tails, driving the quills deep into an attacker's flesh. The quills are armed with tiny barbs. This not only makes them hard to remove but means that, if left inside, they tend to work themselves deeper and deeper into the tissues. The penetrating wounds are intensely painful, frequently become infected, and must be removed by a veterinarian or physician. The usual victim is the family dog, whose snout is the customary receiver of most of the barbs.

The porcupine, *Erithezon dorsatum,* is 18 to 22 inches long, weighs ten to 28 pounds and has a seven- to nine-inch-long mus-

Porcupine

cular tail. These animals are heavy-bodied and have short legs, which gives them a clumsy-looking, slow gait. Porcupines have blunt snouts, furry underparts, tiny ears and small, black eyes. They are excellent climbers and spend much of their time in trees, hence are sometimes called "tree porcupines." They are found across much of Alaska and Canada and in the northern parts of the United States but are not found in the Deep South or Southeast. They are mainly nocturnal and have a distinctive, dink-toed track.

Porcupines produce one offspring a year, the young reaching sexual maturity by the time it is two. Porcupines are often gregarious, and it is not uncommon to find several of them sharing the same den.

Porcupines damage trees by gnawing off their bark and can raise hob with vegetable gardens, munch on lily pads in fish ponds and gnaw on wooden buildings. They crave salt, as do most mammals, and have been known to gnaw on wooden tool handles that are salty with sweat.

Although they prefer nesting in hollows in the woods, they sometimes will build their dens under buildings. This can be prevented by using metal sheeting or heavy-duty wire mesh to block all potential openings.

Control: Humane live traps can be used to rid yourself of a few local pest porcupines, but make sure you release them far away from your area, since they frequently will range a few miles in search of food. They can also be killed with poisoned baits (see page 17) that are effective against rats. In rural areas, if local laws permit, they can be jacked, using a spotlight and .22 hollow points. If you get them on the ground, they can be directed into a trap or cage with a long-handled rake. But be careful: they can make short-distance quick lunges and swat you with their tails.

RACCOONS

Raccoons, *Procyon lotor,* or "coons" as they are sometimes called, are remarkably adaptable animals that are now reasonably common in built-up suburban areas. These grizzled, gray, furry, nocturnal animals with their bandit's black eye masks and ringed, bushy tails are extremely dexterous and adept at overturning trash cans, tearing open plastic garbage bags and causing enough of a commotion at night to wake even a sound sleeper. If you have dogs, the racket worsens as they respond to the coon's presence by howling and barking. Unfortunately, coons can be ferocious when cornered, and a 35- to 40-pound adult male raccoon can

Raccoon

more than hold his own against even a good-size dog. The resulting veterinary bills or in some cases the death of a beloved pet is a high price to pay so, if your dog trees a raccoon, take the dog inside.

Raccoons are carnivores belonging to the Procyonidae family. Their heads and bodies range from 18 to 28 inches in length and their bushy ringed tails eight to 12. Adults usually weigh from 12 to 35 pounds, but a rare adult male may reach almost 50 pounds. Their thick fur is grayish-brown and their tails have alternating rings of yellow-white and black. Raccoon footprints are distinctive: their claws are not retractile, and, since they are flat-footed, the marks of all five toes on both front and hind limbs are clearly visible. Their slender front paws are excellent for grasping and manipulating objects and they are good climbers.

A single litter of about four raccoon pups is born in the spring, but the young don't leave the nest for the next two months, after which they can be seen in line behind their mother on her foraging forays. I find these animals quite charming and suggest that simple exclusion techniques are the best to follow.

In the past two years, a new raccoon problem has emerged in some parts of the country as rabies seemed to reach epidemic proportions in some raccoon populations. Until this recent outbreak, raccoons had not been considered a major rabies virus reservoir, but now raccoons around the house pose a potentially lethal hazard to humans. Be particularly careful of raccoons that are not frightened by you, since fearlessness may indicate the animal is rabid. (see Rabies, page 27).

Raccoons are also excellent climbers and not infrequently get into attics or lofts and set up housekeeping. Their scurrying about in your overhead and their occasional shrill cries or whistles disturb a good night's sleep. Most raccoons are also infested with fleas and ticks that might find their way into your living quarters. They are also capable of transmitting a number of infectious diseases besides rabies; hence it is a good idea to deny them access to your attic.

Prevention and Control: To stop raccoons from scattering your trash, use containers with well secured lids and keep these in a rack that prevents toppling. Raccoons will occasionally fish in garden goldfish ponds, devouring some rather expensive pet fish. A wooden frame with chicken wire placed over the pond will keep out fish-hunting marauders such as raccoons and the local pet cats.

Raccoons usually gain access to lofts and attics through openings in the eaves. The best tactic here is to deny access by closing the opening, but first make sure that the animals are out. Moth crystals (naphthalene or paradichlorobenzene) may be effective in repelling the squatters in your attic.

If raccoon problems are serious, you may have to use control measures. Live traps (Havaharts or Tomahawks of appropriate size) are somewhat successful, but these animals are smart and will often avoid these rather expensive contrivances. If you decide on trapping, use fish or fish-flavored cat food as a bait. Be careful letting the raccoon out of the trap, because it is likely to have found the whole experience quite upsetting and be rather aggressive. Leg-hold traps are both dangerous and inhumane. I really can't condone their use. In rural areas that are sparsely inhabited, shooting is effective when used with a spotlight. Make sure, however, to check with your local game control officer, since in some areas raccoons may be protected. Shooting in most suburban areas is a violation of the law. Local dog control officers or SPCA people may be willing to get rid of an occasional raccoon for you. Local pest control companies can also provide such a service but at some expense.

COYOTES

The coyote, *Canis latrans,* is a carnivore belonging to the same family as the domestic dog, foxes and wolves. They live in many parts of the United States and Canada, being particularly numerous in the open ranges and mountainous foothills of the western one-third of the continent. While their usual food is wild rodents, and they help maintain ecological balance, they can become serious pests particularly in sheep herding areas and in areas where the suburban homes encroach on their range. In this discussion, the coyote problem will be limited to the suburban foot-

Coyote

hills of Los Angeles and Southern California where coyotes have frequently attacked family pets and in one recent confirmed case killed a small child in a residential area.

Coyotes look like scrawny, small wolves having a head and body up to three feet long and a bushy tail a foot or more in length. Their weights range from 20 to 50 pounds. Their fur is gray to rusty gray, the legs, feet and ears are rust-colored and the undersides, including the throat, are light gray to white. When they run they keep their tail down behind their hind legs. They hunt singly, in pairs or even small packs. They are extremely clever animals, and while not foolhardy, have little apparent fear of humans, and often announce their presence at night with a series of high-pitched yaps.

Control: Despite attempts to control this wily predator by trapping, shooting, and poisoning, the coyote still flourishes. Control in residential areas should be left to local authorities but homeowners in coyote country can prevent harm to their children by enclosing their yards with tall fences (over six feet) since coyotes are remarkably good jumpers. It's also a good idea not to let your small pets run free.

SKUNKS

Widely distributed throughout the United States and Canada, including many suburban parks of major cities, are the handsome little carnivores known as "woods pussies," "polecats" or skunks. About the size of an ordinary housecat, these animals have relatively little fear of humans and will often decide to move in and set up housekeeping in or under buildings, under porches or under mobile homes, even in densely built-up neighborhoods. They are scavengers and frequently will raid garbage cans. The main problem with skunks is their smell. Their defensive armament is awesome, and they can, if disturbed, eject a foul-smelling secretion from their scent glands. This potent liquid is aimed with considerable accuracy and can hit a target at a distance of ten feet. The fetid, sulfurous liquid is nauseating, and if it hits the eyes produces a severe burning sensation and even temporary blindness. Children and pets are the usual victims, with the most frequent victim being the family dog. Dogs are curious and very defensive of their own territory, unfortunately, they usually come out losers in duels with skunks. The aftermath of such confrontations is a rather uncomfortable and very smelly pet, an unhappy homeowner and a veterinary bill for having the dog washed.

There are five species of skunks in the United States and Canada, but only three are sometimes pests. The commonest of these is the widely distributed striped skunk, *Mephitis mephitis*. These animals have bodies from 13 to 18 inches long and seven- to ten-inch-long bushy tails; they weigh from six to ten pounds. Their fur is mostly black, with a narrow white stripe on the middle of the forehead and a broad white stripe starting at the nape of the neck that divides into two stripes just behind the shoulders and continues to the rump. The tail is usually black and often terminates in a white tip.

Spotted skunks, *Spilogale gracilis* (the western species) and *Spilogale putorius* (the eastern species) are found everywhere except on the East Coast and around the Great Lakes. These animals are smaller than their striped cousins but are better climbers and have been known to get into houses through open, unscreened windows. The head and body ranges from nine to 14 inches in length, and the tail from four and one-half to nine inches. They weigh only one to two pounds but are still potent stinkers. Spotted skunks are black with white spots or streaks on their

Striped Skunk

foreheads and under each ear. They have broken white stripes or spots on their necks, backs and sides, and their bushy tails have conspicuous white tips.

Skunks usually produce one litter of four to seven kittens per year, most often in May, June or July. They become sexually mature in about a year and are relatively long-lived (up to 12 years in captivity). A number of pet stores sell surgically de-scented skunks as house pets. They certainly are novel, and the few people I've spoken to who have owned pet skunks compared them to affectionate cats.

Spotted Skunk

Skunks' nests are usually burrows, but they settle readily under buildings, and often more than one family will occupy a den. Sometimes they will leave a bit of telltale odor—but it's not the odor that's the major problem. Skunks are one of the major reservoirs of rabies virus (see page 27), and all skunk bites—indeed any wild animal bite—requires the immediate attention of a physician since, if untreated, the bite of a rabid animal is about 100 percent lethal. Be particularly careful of skunks that wander about in a fearless and lethargic manner during the daylight hours, because such behavior may be an indication that the animal is rabid.

Prevention and Control: The best way to deal with skunks under buildings, porches and trailers is to prevent their entry physically. Block up all potential entries to their dens but first make sure they are out. If they are under a mobile home up on blocks or under a porch, they can be driven out by using floodlights trained on the den area—they have an aversion to light. Or you can use the same tactics the skunks would use—stink them out. Sprinkle one or two pounds of mothballs or mothflakes into their den area or use a shallow baking tray filled with household ammonia to drive them out. Then block reentry.

Skunks are easy to trap with humane live traps, but sometimes handling the trapped skunk results in being sprayed at close range. Wear goggles to protect your eyes from the irritant spray if you want to attempt live trapping. Above all, skunks are not aggressive, and if you don't move in a threatening way the chances are that they will not unlimber their artillery. Shooting and use of poisoned bait is not recommended unless there is a rabies epidemic. The best advice I can give is deny them nesting places and keep your dogs in at night.

How to Deodorize Skunk-Sprayed Dogs

There are a number of myths about ways to neutralize skunk odor. One favorite is to wash the dog in tomato juice. I tried this once and ended up with a very smelly red poodle that also smelled like tomatoes. Skunk odor is very persistent, but one teaspoonful of neutroleum alpha per gallon of water will provide a reasonably effective odor-neutralizing bath for sprayed dogs (or people). It can also be used to wash down skunk-contaminated floors and furnishings; contaminated soil can be deodorized by liberal application of the same solution. For this purpose, make the solution three times more concentrated than the dog bath.

OPOSSUMS

The opossum is the only representative of the ancient mammalian order Marsupalia in the United States. Marsupials give birth to tiny young that complete their development while attached to a nipple in the pouch on the belly side of the female. Commonly called 'possums, they are a favored gastronomic delicacy among some parts of the southern population but elsewhere are considered to be no more than occasional minor nuisances. They are hardy creatures and survive well in built-up woodsy suburban areas as well as rural areas. These harmless, nocturnal, cat-sized animals frequently will set up temporary dens under mobile homes, buildings, porches, crawl spaces and even attics. They are omnivorous and often will raid garbage cans and tear open plastic trash bags in the search for their evening meals. This not only leaves a mess to be cleaned up but also gets the neighborhood dogs aroused, producing a considerable amount of noise. Usually they occupy these dens for only a short period of time and then move on. As with most wild animals that set up housekeeping in gardens and around human dwelling places, there is a potential problem with disease transmission and with the introduction of a variety of ectoparasites (ticks, mites, fleas, etc.).

The 'possum, *Didelphis virginiana,* is an adaptable animal whose distribution was originally limited to the eastern part of the country but has now spread all the way to the west coast. It has a body length of 15 to 20 inches and a long, prehensile, rat-like tail that ranges from nine to 13 inches in length. Adults weigh from four to 12 pounds, have squat, heavy bodies and short legs and are covered with coarse, long gray fur. This dense, fine-grade fur is prized as a warm lining material for coats. The 'possum has

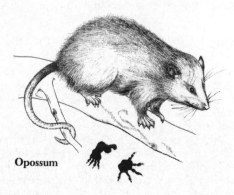

Opossum

a long, pointy nose, a white face, and small, jet black, paper thin ears. 'Possums have five spreading toes which, along with their prehensile tails, make them excellent climbers.

Opossums produce one litter of from four to eight young per year. Initially, the young stay in the pouch and, when they do leave it, they ride the mother's back for a while before they are large enough to accompany her on her nocturnal meanderings. 'Possums don't move very fast and seem rather sluggish. When confronted by a dog they will scoot up the nearest tree, leaving a noisy, barking pet to awaken the neighborhood. By and large, these animals are nothing more than a rare nuisance and don't require control, but if they do become a problem the best solution is to deny them access to den areas around human habitations.

Prevention and Control: Use the same tactics and strategies suggested for "building out" raccoons and skunks. Some burrows can be treated with mothballs or moth crystals, which act as repellents because of their irritating vapors.

In some rural areas, opossums may become pests around poultry houses and can be live-trapped, netted, hunted with dogs or shot, but best check with the local game department first, as there may be seasonal protection for these animals.

ARMADILLOS

The armadillo is a tropical, armored, primitive mammal that has extended its range to include Texas, Oklahoma, Arkansas and Kansas. It is also spreading eastward to Georgia and South Carolina, possibly making its way there from populations introduced into Florida. Armadillos are burrowing animals that can damage lawns and gardens as they root in search of food. In addition to tearing up the landscape, they sometimes burrow under buildings and leave their distinctive, unpleasant odor behind. These harmless animals are really not serious pests, and in some parts of the country armadillo meat is a prized gourmet food.

The armadillo, *Diasypus novemiticus,* has a body up to 17 inches long, has an equally long tail, and weighs eight to 16 pounds. Its horny armor plates protect most of its body except for the tip of its snout, prominent ears and undersides. Small, scattered hairs project from between the armor plates. When disturbed, it rolls into a protective ball. Armadillos, for all their armor, are surprisingly quick. There are even armadillo races. These nocturnal grubbers can swim and are excellent diggers. They produce one litter of four offspring per year; their population is on the rise.

Armadillo

Prevention and Control: If armadillos become a pest in the garden, fence in the plot. Lawn grubbing can be stopped by depleting the lawn animals such as grubs (see page 169); burrows can be fumigated (see page 18). Live-trapping is effective, and in areas where firearm control permits, armadillos can be jacked, using a spotlight and .22 rifle.

BIRDS ON AND AROUND HOMES

A number of species of birds have come to depend on human habitations as nesting and feeding areas. While most should be regarded as welcome guests that are allies in the fight against insect pests, a few, the so-called "trash birds"—house sparrows, pigeons, starlings and seagulls—can become the source of annoyance and cause problems. These problems include noise, messy nests, foul droppings and, in some cases parasitic, bloodsucking pests and disease organisms for which these birds are sources.

STARLINGS

The most numerous bird in the country is the robust, short-tailed iridescent blue-black, white-speckled, nine-inch-long, spike-beaked European starling, *Sturnus vulgaris*. These noisy inhabitants of urban, suburban and rural areas were introduced into the United States on March 5, 1890, by a well-to-do New York druggist, Eugene Scheifflin, who wanted to introduce Americans to all the birds written about by Shakespeare. The ninety starlings he released in Central Park proliferated and spread to all parts of the country.

Early in the morning, small formations of a dozen or more individuals, wheeling in tight formation, may arrive in your yard, noisily whistling and screeching like un-oiled shopping cart wheels. This description really isn't totally fair, because starlings, in addition to their own harsh, metallic, sharply descending whistles, frequently imitate the songs of some of their more melodious cousins. Once the flock has descended, they begin their march in search of food. They are omnivorous and can riddle your lawn as they deftly pick grubs and worms from the soil with their long, pointed beaks. In this scenario, they may actually be of considerable benefit in getting rid of insect pests while simultaneously aerating and fertilizing your lawn.

Starlings are very sophisticated birds and are extremely wary and adaptable. They are just as adept at depleting your berry bushes and picking over your trash as they are at eating grubs in your lawn. Some people dislike starlings because their noisy early squawking wakes them up. Others get very unhappy with their copious droppings, often splattered on windshields and hoods of freshly washed cars, but the main problem with starlings comes when they set up their nests in or on your home. Or, worse yet, when small flocks merge into gigantic flocks and set up huge communal roosts (thousands of birds) on the I-beams of bridges, in church belfries or in the trees of city parks. As they wheel into their roosts at night and when they depart en masse to their feeding grounds, commuters are bombarded with a barrage of starling droppings, a form of behavior not guaranteed to win them endearment in any form.

Starling

Nests on the home may pose a problem, since the accumulation of their smelly droppings may cause histoplasmosis, and some of the starlings' external parasites (fleas, mites and ticks) can attack humans if they get into the house. Starlings around airports are a more serious problem, and on October 6, 1960, starlings sucked into the engine of a turbo-prop Lockheed Electra leaving Boston's Logan Airport caused the plane to crash, killing 62 people.

Starlings are very aggressive birds and will often build their nests in other birds' nesting hollows. They mate in the spring and lay four or five green-blue eggs from one to three times a year. Within a month, the fledglings are out flying with the flock.

Control: The best tactic in getting rid of pest starlings is to deny them nest sites around or on buildings. Thus, whenever you see starlings nesting, block off those ledges, nooks and crannies with wire or nylon netting. In the case of large communal rookeries, contact your local health department, since such roosts are really a community problem. Use of fireworks, recordings of starling distress calls, rubber snakes, etc. have only limited success. Wire repellent spikes and sticky repellents do have some success. Use of poison baits or toxic perches may be effective in dealing with small numbers of birds, but may also kill non-pest species or neighbors' pets. And this is not good for neighborhood public relations. There is also a problem associated with dead birds littering neighbors' lawns. Shooting with airguns or .22 shorts has obvious risks and, being very wary, starlings will allow only a single shot before they fly off. In summary, deny roosting sites, wash your dropping-splattered car and don't get too upset with starlings, since they are interesting and do some good.

PIGEONS

The common wild pigeon, *Columba livia,* is descended from the wild rock doves of Europe and Asia and from antiquity has had a close association with humans. They have been raised and kept as a food source; they have been kept as pets and as carriers of messages, and have been used in games by pigeon fanciers. Domesticated pigeons arrived with some of the earliest settlers of North America, and the offspring of these birds, mostly as wild pigeons, have flourished in every urban and suburban part of the United States and Canada—where they have become a considerable nuisance around homes and parks.

They are very adaptable birds and have little fear of man. At

Pigeon

times, their strutting, billings and courtship behaviors are quite interesting, and some pigeons are really quite handsome. They frequently become tame enough to eat from your hand. They become troublesome pests only when they roost or build nests on your home's eaves and rafters. Their nests may clog drainpipes and chimneys, and their noisy cooing may disrupt sleep. Their close association with the house may also introduce some of their parasites (mites, ticks, etc.) into the home, where they can be a temporary problem. Finally, their messy, smelly droppings, randomly splattered on human belongings, not only are unattractive, but may transmit diseases, e.g., histoplasmosis, encephalitis aspergillosis, food poisoning and toxoplasmosis, to humans. Pigeons' droppings are quite acidic and not only are messy but cause deterioration of buildings and statues. Indeed, if statues could speak, their main complaint would undoubtedly be pigeons.

Unlike most humans, pigeons are monogamous through their long life spans—fifteen or more years. While they breed all year, the peak reproductive period is spring and summer. They lay one or two eggs, which hatch in 18 days. The hatchlings fly by the time they are six weeks old. Fortunately, these fledglings have a high mortality.

Control: The first step in getting rid of pigeons on homes is to deny them roosting and nesting sites by screening off exposed nesting sites with rustproof wire or nylon netting, both of which are available commercially. Other strategies are: installing (rather expensive) high-voltage wires of low amperage, mounting sharp, pointed wires or spikes that prevent landing, emplacing plastic or metal flashing at angles of 45° in nesting areas, applying commercially available bird repellent adhesives or jellies to nesting areas or simply discouraging nesting by removing nests every two weeks. All these procedures have some risk, since nesting areas are almost always so high that they can be reached only by means of a very long ladder. It is hardly worth risking life and limb, so, unless you are agile, it might be a good idea to contact a licensed and bonded pest control operator.

Baited live-catch traps that are available commercially are highly effective, but the problem of ridding oneself of the trapped pigeons remains. Driving them out into the country and releasing them isn't much good; since these birds are marvelous navigators and will be back at your house almost as soon as you are.

Shooting is also effective, but in most urban and suburban areas this is illegal, and a miss or ricochet may endanger people. Poison baits are very effective in dealing with pigeons, but check with your local agricultural Extension Service regarding the legality of using strychnine, the bait poison of choice. There are three problems with poison baits: (1) you may poison non-target species; (2) the dead birds have to be picked up and (3) if your kids or spouse gets mad at you, you may end up as the recipient of strychnine. Availability of poison seems to be a factor in both homicides and suicides . . . and then of course if there are young children, there is also the problem of accidental ingestion, since very young children (under five) will put almost anything into their mouths.

A new technology, the poison perch, is highly effective. This consists of a tube that acts as a roost. It is filled with a wick containing endrin or fenthion, both effective toxins that are absorbed through the skin of the bottom of the feet of the pigeon. These perches, placed in carefully selected pigeon roosting areas, minimize risk of poisoning non-target birds or pets. However, they are expensive, and there is the problem of dead birds littering the neighborhood. Different states have established their own regulations about which toxins may be used in poison perches and in poisoned baits. Check with your local Agricultural Extension Office for what control measures are permissible.

ENGLISH SPARROWS

The house sparrow or English sparrow, *Passer domesticus*, is actually a weaver finch that is extremely fond of human habitations. The Duke of Wellington, victor of Waterloo, was once asked by the Prince Consort what to do about sparrows that had gotten into the Crystal Palace, a huge, glass-domed exhibition hall. Wellington answered: "Sparrow hawks." Unfortunately, his solution is not practical for the home owner plagued with these smelly, prolific, persistent urban and suburban "trash birds."

The English sparrow is a small (from five to six inches long), grayish-brown, stocky, highly social bird. The male has a black throat, a gray crown and a brown streak extending from the eyes to the neck. These prolific pests may produce as many as five broods of up to eight offspring each per year. Fortunately, more than half of these die before attaining maturity; otherwise, we would be inundated with house sparrows. Their messy nests, built of grass, string and paper, can be anchored onto any protected, elevated place on a house, and they are adept at finding openings and building nests in attics. They not only deface homes with their droppings but may also transmit a number of diseases, e.g., encephalitis, salmonellosis, psittacosis and others, and some of their parasites (fleas, ticks, sparrow bugs and mites) may find their way into the house and become temporary itch-causing, disease-transmitting pests.

English Sparrow

Control: Restrict entrance to homes by plugging holes and screening air vents! After bird-proofing, begin the systematic destruction of nests at two-week intervals. But you have to keep it up to discourage the sparrows. Hard-to-reach nests can be hosed down or pulled off with a long, hooked bamboo pole. House sparrows love English ivy as a nesting site, and these nests are tough to find and remove. Hence, the solution here is to cut down the vines. Otherwise use anti-nesting and -roosting strategies and trapping techniques to those described for starlings and pigeons (see pages 42 and 44).

MISCELLANEOUS AND OCCASIONAL BIRD PESTS

SEAGULLS

These large, handsome shore birds have become a major pest in some coastal areas of the United States and Canada. They are relatively fearless and extremely adaptable. With their powerful beaks, they are adept at ripping open plastic garbage bags, making an enormous mess in the process. They are particularly annoying around dumps, where vast squadrons congregate to "pig out" on humans' swill. As the well-filled birds swoop down, they let fly with a veritable blizzard of white-and-brown, smelly droppings that are extremely unappetizing when splattered on your car, house eaves or—worse yet—you. They also have the nasty habit of picking up live clams and dashing them on hard surfaces to break open the shells to get at the tasty morsels inside. One summer a friend of mine had to sweep off his tennis court every morning to remove the shells and glop. I must say they are certainly far more accurate as bombardiers than were my colleagues in the 8th Air Force during World War II, since the seagulls seldom miss their targets and we seldom hit ours.

Control: None feasible; just resign yourself to the inevitable splat.

WOODPECKERS

Woodpeckers and flickers are equipped with long, tough beaks and extensible tongues which they use to extract insect larvae from tree burrows. With their powerful, hammer-like blows, they are adept at chiseling into wood and herein lies the problem: for reasons unknown, they sometimes attack wooden parts of houses. If they decide to drill into your external woodwork they can be very persistent, returning several times in a single day. The holes drilled by woodpeckers not only disfigure the building in question, but unless the damage is repaired may allow wood-damaging fungi an entrance. The noise of a woodpecker's hammering can be a considerable nuisance to anyone inside the house. What I've never understood is why woodpeckers don't get headaches from all that pounding.

Control: Woodpeckers are protected by law; hence all you can do is fasten protective sheeting onto the area under attack or suspend nylon netting in such a way that they can't get at that part of the building favored by them. Humane live traps baited with suet may also be effective, but you will then have to chauffeur the bird to a new area and release it.

BARN SWALLOWS AND CHIMNEY SWIFTS

Both swallows and swifts are small, agile aerial hunters, with the swifts flying high patrol and the swallows sweeping in low to gobble up enormous numbers of gnats, flies, mosquitoes and other flying insects. Since they do so much good and are really wondrous, attractive, acrobatic fliers, I don't think it is really fair to classify them as occasional minor pests. The problem comes only when they make their nests on your home or outbuildings. The monogamous swifts return to the same nest year after year, adding more and more material to their nests high up on buildings in sheltered areas, including chimneys, which may eventually become clogged. Barn swallows nest in rafters on the insides of buildings (barns and sheds) where they have continuous access to the outside. The nests are messy, cup-shaped structures made of mud, dry grass and other materials. These nests, while hardly aesthetic structures, should be tolerated until the swallows depart for the winter since unlike the swifts, swallows don't return to the same nest. Frankly speaking, anyone who would be annoyed by swallows shouldn't have this book.

BIRDS AS ALLIES

In the previous section of this book we listed a number of "trash birds" that may cause problems for the homeowner, but these inconveniences are minor when compared with the benefit we derive from most of our feathered friends. Birds are of immense value in controlling many insect pests that damage your garden, invade your home and cause you painful or itchy bites and stings. Even nominally fruit- and seed-eating birds, when feeding their young, dispatch large numbers of insects; indeed, their gastro-nomic feats are quite prodigious.

A single house wren, a common inhabitant of human living areas, will bring up to 500 caterpillars to its nestlings in one day. Flycatchers average fifty insects per hour. Yellow-throated vireos have been observed to eat 120 cankerworms in an hour; a chick-adee feeding its young will bring in 70 insects an hour all day long. A house wren, observed for 15 consecutive hours, fed its young 1,217 insects. Even birds commonly considered pests pay their dues. A flock of noisy starlings patrolling your lawn will not only aerate the soil but also will gobble up hundreds of grubs per hour.

Ideally, you want both numbers and variety of birds in your living area, since each species will occupy its own unique feeding niche. Swifts, swallows and nighthawks are aerial hunters that hit on such flying insects as mosquitoes, gnats and biting midges. Other birds are ground feeders, some in open meadows, and others on the forest floor. Woodpeckers, with their long, sharp and powerful beaks, pick wood borers and grubs from otherwise inaccessible burrows. Barn owls and other small owls that may set up messy housekeeping in your barn, shed or attic help keep the rodent population down. Some birds prey on pests on the undersides of leaves, while others prey on insects that feed on leaf tops. Even sparrows and doves, which are almost exclusively seed feeders, get the majority of their food from common weeds that would otherwise clutter up your garden. Certainly weed control by birds is preferable to chemical herbicides. Besides, these toxic chemicals don't fertilize your lawn the way birds do with their droppings.

ATTRACTING BIRDS

In order to forge an alliance with our insect-eating feathered friends, it is most helpful to attract them to your home site by feeding them through the winter. Although birds are well adapted to tolerate the cold, they may need help with food. Starting in the middle of October onward through March, when their natural food is most scarce, you should continuously stock your bird-feeder.

There are many types of feeders, any of which will attract birds. The simplest feeder is an open platform feeder mounted on a post in a sunny place near shrubs or trees. The platform should have one-inch-tall edges to prevent the feed from blowing away, and a small, forked, upright branch should be mounted at the edge of the platform as a landing site for the incoming birds. There should be several small (⅛-inch) holes in the platform for drainage. Covered feeders may be even better. There are commercially available, finished feeders and kits, but if you wish you can make your own from scratch. To do so, build a platform with two eight- to ten-inch-tall wooden sides and front and back open. The top can be glass or lucite so that the birds can see the feed. Mount a round stick or dowel across the front to act as a perch. Remember to be consistent in your feeding if you want the birds around when the insect season arrives.

Bird feed mixtures are widely available and not terribly expensive. Your feeder may also attract squirrels and chipmunks, but there are some commercially available squirrel-resistant feeders. In addition to providing wild birds with mixtures of seeds, you can attract some species with a suet holder. Beef suet (fat), bacon or pork fat is solid when cool. To make a suet feeder, build a pocket of ½-inch wire screen or hardware cloth or even an old Christmas mesh stocking, fill it with suet and nail it to a branch or post. Peanut butter will do as a suet substitute. Some people mix the suet or peanut butter with crumbs or dry cereal.

PART II

Invertebrates That Bite or Sting

Insects are arthropods, joint-legged animals, that have a tough external skeleton, three distinct body parts, (head, thorax and abdomen), and three pairs of legs; most have wings. There may be five million species of insects. They range in size from 12 inches in length down to the size of a dust particle, invisible to the naked eye. Insects first appeared on earth about 400 million years ago and have remarkable properties for adapting and surviving. Their tough external skeleton, their flight capabilities, their prodigious reproductive capabilities and their physiology allow them to endure conditions that could and did destroy other animal species. Some insects can survive temperatures as high as 120°F, while others can handle temperatures as low as minus 30°F. Insects can even tolerate levels of radiation that would be lethal for most other animals.

Insect survival is enhanced by a life cycle that goes from the egg through a series of larval stages during which there is a gradual change to the adult form. Some insects undergo metamorphosis, a complete change in form, which includes a pupal stage in a cocoon. If environmental conditions are disadvantageous, they may suspend development at one of these stages until conditions improve. Their reproductive capacity is mind-boggling, with some insects capable of laying hundreds of thousands of eggs per mating. The end result is that there are a trillion billion insects in the world, and their total weight is more than ten times the weight of the earth's human population. Considering that the average insect weighs less than a thousandth of an óunce, their numbers are indeed astounding.

Many insects are beneficial to humans, but many are also serious pests that devastate our forests, destroy our crops and decimate stored foods. Other insects are carriers of diseases that have caused millions of human and animal deaths; some stinging and biting insects, though rarely lethal, inflict painful bites, raise itchy wheals on our skin or disturb our sleep. This section will con-

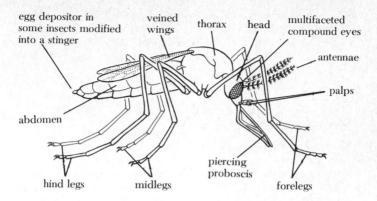

centrate on those families and species that are frequently encountered and either bite or sting. It will include sections on: insecticides, insect repellents, symptoms of stings and first aid.

Insects are grouped into orders based on common characteristics. The orders are further broken down into families, genera and species. Sometimes the ability to correctly identify the insect has significant bearing on the treatment; hence it is essential to know either the scientific or common name of these insect pests.

There are three major anatomical parts of an insect's body: the head, the mid-part or thorax and the segmented abdomen. The head bears a pair of remarkable multifaceted compound eyes that give the insect an almost 360° field of vision. It also is equipped with a pair of chemical sensing antennae which allow the insect to find its prey and mate but also responds to repellents. Last but not least, it is equipped with a variety of mouth parts. The insect shown here, the pestiferous mosquito, has piercing-sucking mouth parts. All bloodsucking insects have some such mouth parts, and when they attack they insert these mouth parts into the skin and inject their saliva to promote the flow of blood. The saliva contains many substances including enzymes, toxins and blood vessel dilators. The initial response may or may not be painful, but the subsequent response produces an itchy area of inflammation. Some humans who are sensitized to the insect may experience a life-threatening allergic reaction and can go into shock and die if not treated.

The mid-portion, the thorax, bears three pairs of jointed legs (spiders, ticks, mites and scorpions have four pairs) and usually one or two pairs of flight wings, although some insects are wingless (e.g., fleas). The hind portion of the body, the abdomen, is segmented and in sucking insects after a meal of blood can become quite distended. Female insects have specialized egg depositing appendages at the end of the abdomen. This organ has been modified in stinging insects into a venom gland and a pointed, sharp, hollow stinger.

THE HYMENOPTERA: BEES, WASPS, HORNETS AND ANTS

The hymenoptera is an order of insects whose members have two pairs of membranous wings and one pair of antennae, and whose female ovipositor has evolved into a very potent stinging apparatus. Included in this order are several species that can inflict extremely painful stings: wasps, hornets, bees, velvet ants and the vicious fire ants and harvester ants. Several million people are stung every year, and while in most cases such stings are merely a painful nuisance with no prolonged effects, there are some people who when stung experience life-threatening medical emergencies or death. According to the Bureau of Vital Statistics, more than 30 deaths per year are due to hymenopteran stings, and of 460 deaths in the United States attributed to venomous animals between 1950 and 1959, more than half were caused by hymenoptera. In the following section the more common members of this order will be reviewed.

BEES

Bees are among the world's most beneficial insects and are the most important pollinators of plants from the Arctic to the Equator. Bees also provide humans with honey and beeswax, but they can inflict painful and sometimes lethal stings. Stinging bees all belong to the family Apidae, which includes both solitary and social species. Bees generally are covered by fine feathery hairs, particularly on the legs and abdomen. While bees have biting jaws, they don't use them for feeding, since they get most of their nutrients by lapping up flower nectar. Most common species of bees have yellow and black stripes on their abdomens and are most active in the months in which plants flower.

The Stinging Apparatus

The bee's stinger is retractible and hidden in a sheath inside the abdomen until it is brought out just as the bee stings. The stinger is about 1/10 of an inch long and is composed of three parts, which surround the poison canal. One part, the director, guides two lancets which move up and down by muscular action. All three parts are barbed and are driven into the flesh. Poison glands in the abdomen are squeezed by the up-and-down movements of the lancets; thus the venom is forced into the wound. The barbed lancets remain firmly imbedded in the skin; when the bee flies or is swatted off, the director, lancets and poison gland are left behind, still injecting poison. Any attempt to pull the sting apparatus out will cause greater injection of venom; hence the stinger must be scraped off with a knife or fingernail (see first aid section).

Bee venom contains several biologically active toxic compounds, most of which were described in the Introduction to hymenoptera, but one deserves particular mention—MCD, a peptide that releases the body's own histamine. MCD is the chief product of bee venom and produces pain, dilation of blood vessels and other allergic reactions. Another toxin, Apamin, is a nerve poison which can cause muscle spasms and convulsions.

THE SOLITARY BEES

Solitary bees, as their name implies, live alone. Each female builds a nest in a particular habitat, provisions it, lays her eggs and leaves the larvae to develop on their own. The Large Carpenter Bee hollows out tunnels in wood, usually in the pithy stems of bushes. She is about an inch long, is robust like a bumblebee and has a naked or hairless abdomen. Her smaller cousin, the Small Carpenter Bee, is only 1/5 of an inch long, is dark blue-green and has similar tunneling habits. Other solitary bees include the Mason Bee, the Miner Bee and the Leafcutter Bee. All have stingers and can trigger allergic reactions in sensitive individuals.

THE SOCIAL BEES

The common honeybees, *Apis mellifera,* are true social bees. They are found in the wild and have been domesticated and bred for their pollinating skills and for their production of honey and beeswax. All of these commercial bees in the United States and Canada were imported. They comprise several races, with the gold Italian and the black and gray Caucasian races making up the majority of bees in the United States. These bees are responsible for the most stings and deaths. In one recent five-year study, Hutchins reported that, of 215 reported deaths caused by venomous animals, 52 were caused by bee stings.

Honeybees are highly social insects whose colonies consist of an egg-laying queen bee, many male drones and very large numbers (thousands) of sterile female worker bees, who care for the

Honeybee

Bumblebee

queen, gather food, and care for the young. Honeybees can release "alarm odors" when stinging, and this may attract other bees to the odor-marked victim, thus producing multiple stings. This alarm odor may also increase aggressiveness. Alarm odors may also account for the aggressive behavior and multiple stings seen when a hive is disturbed.

BUMBLEBEES

Bumblebees are considered to be the most primitive of the social bees. Their colonies lack much of the structure and highly evolved behavior of honeybees; however, like honeybees, bumblebees are diurnal plant feeders and important pollinators of crops. A typical colony consists of at least one queen, several males and numerous workers. Only young fertilized queens survive the winter to establish new colonies the next spring. The nests are normally located deep in undisturbed ground, like fence rows, and are supplied with a mixture of pollen and honey. During late summer, a colony usually contains between 100 and 500 bees. Although bumblebees are two to three times larger than honeybees, they are neither as aggressive nor as abundant as honeybees, and therefore not as dangerous, but they do produce a fearsome loud buzzing noise as they fly and thus wage psychological warfare.

WASPS AND HORNETS

Most wasps have slender bodies composed of three distinct body parts. The mid-part, the thorax, bears two pairs of wings, the anterior pair being larger than the rear pair. One family, the Mutillidae or velvet ants, are covered with bright red or orange velvety hairs. The females are wingless. Another family, the Vespidae, have tiny smooth hairs and are typically yellow and black, white and black, dark brown or steely blue. The wasp abdomen is egg-shaped and attached to the thorax by a very slender stalk, hence the term "wasp-waisted" for women with very thin waists. Through this inelastic narrow stalk run the digestive tube, blood vessels and nerves; because of this constriction, wasps can eat only liquid foods.

Only female wasps and hornets have stingers, and these are generally larger and sharper than bee stingers. Moreover, they lack barbs; thus the wasp can voluntarily remove its stinger and sting again. Venom is injected by contraction of muscles surrounding the venom sac. Wasps also have strong jaws—mandibles—and are capable of both biting and stinging. Since many wasps are scavengers, the bite can be a source of infection.

Wasps, like bees, can be divided into solitary and social types.

Solitary Wasps

Among the largest and most beneficial wasps are the solitary forms that prey on a variety of insect pests. These solitary hunters paralyze their prey and carry them to their nests, where the eggs provide the developing wasp larvae with enough food to carry them through development. The largest of the solitary hunter wasps is the Tarantula Hawk, of Mexico and southwest United States. This large, steely-blue-black predator feeds on tarantulas. There are mason or potter wasps that make jug-shaped mud nests, digger wasps that make nests in the ground and the common, long-waisted, brown mud dauber, a wasp that sticks mud nests on walls, rocks or other protected sunny areas. These nests are tubular cells, and the mud dauber fills from six to 20 tubes with up to 20 paralyzed immature spiders to feed her developing young, which emerge as adult wasps in about three weeks. Another common large hunter wasp is the cicada killer. Adults emerge in early summer after overwintering as resting larvae. The adult female mates, and then digs a long tunnel in the ground ending in an oval cell. Then she goes hunting, stinging and paralyzing one or two cicadas which she drags into the nest. She lays a single egg in the chamber, seals it and flies off to dig another cell. A completed nest may contain up to 16 cells. These large solitary wasps are highly selective in their prey and will only sting humans by accident. Not so with their more aggressive bretheren, the social wasps.

The Social Wasps

There are four common species of social wasps that humans encounter frequently. These are the yellow jackets, bald-faced hornets, European hornets and the polistes wasps. These social wasps are generally considered dangerous because of their large num-

White-faced Hornet

Mud Dauber

Yellow Jacket

Paper Wasp

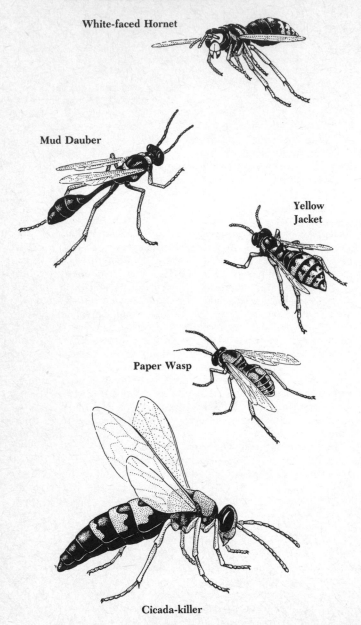

Cicada-killer

bers, nesting habits and very aggressive behavior. They all chew up fragments of wood and leaves, which they mix with their own saliva to make a form of paper used in constructing their nests. These nests may be built underground or in the open, attached to branches or projecting eaves of cabins or houses. The nests have several layers of six-sided paper chambers covered by several layers of paper. Usually there is only one entrance-exit opening. Open nests are usually spherical. Queen wasps lay eggs in each cell, and the larvae are fed on insects collected by worker wasps. If their nests are disturbed, social wasps pour out in large numbers to attack the intruder.

Yellow Jackets *(Vespula pennsylvanica)*

Many a barbecue or picnic has been disrupted by these very aggressive stinging pests called yellow jackets. These black-faced wasps have yellow and black stripes and are about ⅔ of an inch in length. While some make paper nests that hang from shrubs or trees, others make their nests in hollows of trees, walls or even in gopher holes. They are attracted to sweet drinks, food and bright colors, are very persistent and bold and have nasty dispositions. They bite and sting without much provocation and can be downright fearsome if their nest is disturbed. A yellow jacket queen can produce 1500 eggs per laying, and these mature into workers in about 30 days, whereupon the queen starts another brood. Thus, a large nest may house a few thousand yellow jackets.

Hornets

The other dangerous paper wasp found in North America is the bald-faced hornet, *Vespula maculata,* and its smaller East Coast cousin, the European hornet. These large, bald-faced, black and yellow-white hornets build football-shaped nests that hang from trees. Such nests are often encountered at the forest edge. The European hornet nests in hollow trees or on the sides of cabins. Unlike yellow jackets, hornets usually leave people alone; however, if their nest is disturbed, they become very aggressive and attack the intruder, inflicting multiple painful stings. A large nest may contain two to three thousand hornets.

Polistes Wasps

Polistes are long, slender, reddish-brown wasps which make their nests in cabins, summer houses and attics, where they winter over. They are not as prolific as the yellow jackets or hornets, but a polistes nest may contain several hundred larvae which hibernate over winter and emerge from their honeycomb paper nests inside the house. The polistes are more of a nuisance than a threat, since they are relatively mild mannered, sting only if provoked and when they do sting it is not as painful as the stings of yellow jackets or hornets.

Velvet Ants (family Mutillidae)

The velvet ants of the family Mutillidae are not ants at all, but are furry solitary wasps. The larger males have wings while the female is wingless. However, she does have a formidable stinger, with which she can inflict very painful stings. Velvet ants are also called cow killers, mule killers or woolly ants. They are from about half an inch to an inch long and are covered with bright orange, red or yellow hairs. They have very thick external skeletons, which provide protection as they run about in search of egg-laying sites. They are most common in the southern and western United States, but one species is commonly found on sandy beaches of Lake Erie during the summer months and causes barefoot bathers considerable pain. They normally sting only when stepped upon or touched by accident.

ANTS THAT BITE AND STING

Fire Ants

In 1918, the South American black fire ant, *Solenopsis richteri,* established a beachhead in the United States, and in 1940, its cousin, the red fire ant, *Solenopsis invicta,* was introduced. Now these ferocious, tiny pests are well established in nine southern states, infesting over 150 million acres. They are highly industrious, well disciplined, social insects that bite and sting. They are tiny—only one-fifth of an inch long—but very aggressive pests nonetheless. Often they latch on with their jaws and pivot, stinging repeatedly. Their nests are usually made in the ground and are easily identified as elevated earthen mounds 18 to 36 inches in diameter and three to 36 inches high, surrounded by undisturbed vegetation. Their nests are also found under boards and stones, in rotten wood and in concrete work—particularly around hearths. These ants are extraordinarily prolific, and a single acre can hold more than a hundred nests, each containing as many as 25,000 workers. If their nest is disturbed, they suddenly pour out of the entrance by the thousands and attack with ferocity. The suddenness of this massed assault may be triggered by some chemical alarm odor, and the end result is that a victim may receive several thousand stings. These immigrant invaders also cause sig-

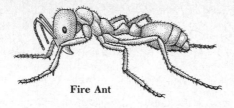

Fire Ant

nificant agricultural damage and attack luckless domestic animals and pets.

Fire ants are currently found in the southeastern United States: Alabama, Arkansas, Florida, Georgia, Louisiana, Mississippi, Texas and the Carolinas.

Symptoms: The fire ant toxin is quite potent and produces both immediate and delayed effects. Initially, the sting causes immediate, severe, burning pain that may last several minutes; then the pain subsides and a localized, raised wheal forms, expands and develops into a tiny, fluid-filled, raised blister. Ten hours later, these blisters flatten out and become filled with pus; these swollen, painful pustules may last several days before being absorbed and replaced by scar tissue. There may be residual effects, such as pigmented areas or two- or three-mm raised scars, persisting for weeks. Most fire ant stings are on the legs.

Some people who have a history of allergies and may have been stung previously become sensitized to fire ant venom. These people can have a generalized systemic reaction and may go into anaphylactic shock and die. Some fire ant-caused deaths have been reported.

Control: On rare occasions, fire ants will invade homes, particularly in damaged timbers or in cracked masonry near chimneys, but most often they will build nests in yards. Given their aggressive nature and painful stings, care should be exercised when applying pesticides. Several of the more potent insecticides previously used in fire ant control are now banned. In limited areas in the yard and in the home, residual poisons in the form of sprays, granules or dusts have proven effective when applied around nest entrances. If using a spray outdoors, use copious amounts of fluid to soak in deep. Spray from a distance, because these ants will sally forth and attack when they feel vibrations around their nests. Call your local county or state Agricultural Officer to find out which pesticide is approved in your area.

Harvester Ants

There are several species of stinging harvester ants of the genus *Pogonomyrmex:* the Florida harvester ant, the red harvester ant, the western harvester ant and the California harvester ant. These diurnal, soil-inhabiting ants form large colonies of several thousand individuals. Their nests are only slightly raised mounds in dry, warm or sandy places. Each nest is surrounded by an area completely cleared of vegetation that can be up to ten feet in diameter. These red or black ants are three times larger than fire ants and are equipped with powerful jaws used to grind seed. When their nest is disturbed, they pour out and attack the intruder in waves, stinging viciously. Small animals have been killed, and humans who have experienced multiple stings have had severe generalized reactions.

Harvester Ant

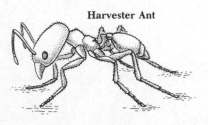

Only the Florida harvester ant, *P. badius,* is found east of the Mississippi throughout the Southeast, while other harvester ants are all found in localized populations west of the Mississippi. While harvester ants don't invade homes, they may set up housekeeping around paths, doorsteps and lawns, where they are a real nuisance. They not only denude lawns but can bite and inflict painful stings which, although not as painful as fire ant stings, are still very unpleasant. If their nests are disturbed, these ants can be quite pugnacious, particularly the California species.

Control: Harvester ants will eat poisoned baits consisting of grain or rye seed placed around the nest. Baits poisoned with mirex or chlordecone are effective, and several new, non-toxic biological control baits are in development. A word of caution about poisoned baits: a number of birds are seed-eaters and should be protected from these poisons. To overcome this problem, put the poisoned seed under a flat rock that is propped up by a small rock.

REACTIONS TO INSECT STINGS

NORMAL REACTIONS

Insect venoms contain a mixture of components including pain-producing substances (serotonin and kinins), blood vessel dilators (histamines), nerve transmitters (acetylcholine), a spreading factor, enzymes, acids, substances that break down blood cells and a convulsion factor. The usual response to a sting of wasp or bee is instant and continuing pain and a small red swelling—a wheal—at the site of the sting. The pain and swelling will usually last a few hours and may leave an area that feels hot and itchy. This will also pass in a few hours, and there is no cause for worry. Stings around eyes, nose and mouth probably cause the most distress, and if stung on the eyelid by a bee, make sure to get the stinger out, since it can cause later complications (see first aid for bees). Some people experience exaggerated local reactions that can last a day or more. Multiple stings, a hundred or more, may introduce enough venom to kill a human, although there are cases of people surviving hundreds of bites. When in Kenya, I saw two people die from multiple stings when they were mobbed by the very aggressive African race of honeybees (the so-called killer bees). Symptoms of such toxic reactions include diarrhea, vomiting, faintness, convulsions, muscle spasms, panic and unconsciousness. The cause of death in such cases is cardiovascular collapse, and as is usually the case, old people or young children are at the greatest risk. A word of caution: when attacked by a swarm, cover your face as much as possible and keep your mouth closed, since stings on the tongue can cause the tongue to swell and you can choke.

ALLERGIC REACTIONS

Most deaths due to bee or wasp stings are due to allergic responses of previously stung hypersensitive individuals. About 30 deaths per year in the United States have been attributed to stinging wasps, hornets or bees, and some medical people feel the number of actual cases may be twice this high. Anyone who has a known allergy to stinging hymenopterans should always carry an anaphylaxis first aid kit when camping away from medical treatment, because the onset of the allergic reaction can be extremely rapid.

The mildest type of allergic reaction to the proteins in hymenoptera venom is a large local swelling around the bite. Sometimes this swelling will extend for several inches upstream from the sting, between the sting area and the heart. Such a response is a warning which may indicate a potential anaphylactic reaction. Anaphylactic reactions involve the immune system and can have widespread effects throughout the body, due to the release of our own histamine and histamine-like substances. Histamine is a potent substance that relaxes the muscle in blood vessels, which then expand. This can cause a severe drop in blood pressure; not enough blood reaches critical organs such as the brain and the victim can collapse, go into a coma and die. Histamine can also cause rapid onset of swelling of the larynx, lips and mouth. The sting victim complains that "my throat is closing." The air passageways can become obstructed, and the lips and fingernails turn blue as insufficient oxygen reaches the lungs. This is a grave sign requiring immediate treatment. Another generalized response by a hypersensitive individual is acute bronchial asthma. The air tubes (bronchi and bronchioles) swell and become obstructed. Onset can be very rapid, and the victim may experience wheezing, difficulty in breathing, shortness of breath, pallor and panic.

The overall anaphylactic response sees a rapid progression of symptoms: the eyes water and become itchy, the inside of the ears may become itchy, the nose may become runny, stuffy and irritated, accompanied by sneezing, coughing, wheezing and shortness of breath; eyes, lips, mouth and fingers may swell to alarming proportions. Hives—large, itchy swollen wheals on the skin—may erupt all over the body. Exhaling may become difficult, and wheezing sounds can be heard during exhalation; lips and finger- and toenails appear purple, blood pressure drops, and there is a loss of consciousness, coma and death. *Anaphylaxis must be treated as a medical emergency.* With adequate treatment, chances of survival are very good, but waste no time—get the sting victim showing the above symptoms to medical help as quickly as possible.

DELAYED REACTIONS TO BEE AND WASP STINGS

Some bee or wasp sting victims show a series of delayed symptoms ten to 14 days after being stung. These symptoms include fever, a general sick feeling, headache, hives, joint pains and lymphadenopathy. Such delayed reactions can lead to anaphylactic shock.

First Aid Treatment for Bee and Wasp Stings

1. If the victim is stung by a bee, the venom sac and stinger will usually remain imbedded in the skin. The sac continues muscular contraction, introducing more venom unless quickly re-

moved. Do not pull out with a pair of tweezers or fingernails. This squeezes in more venom. Scrape off the stinger with a knife blade or fingernail.

2. Wasps, hornets and velvet ants do not leave their stingers, but they are scavengers, and their stingers can cause infection hours or days later. Wash off all stings with soap and water and be alert for signs of infection.

3. Elevate and rest the stung extremity and apply an ice pack to reduce the intensity of the pain and the duration of the swelling. Some medical people think a soak in 10% ammonia solution or application of a baking soda and water paste may relieve the pain and swelling, but these treatments' value may be only psychosomatic.

4. Oral antihistamines may be helpful in exaggerated local reactions, as will topical steroid creams.

5. People with known allergies should wear a medi-alert bracelet or tag.

Prevention and Control: The best way to avoid hymenoptera stings is to prevent human contact with these insects. This is especially important for persons hypersensitive to bee venom. Some preventive steps are:

1. Avoid outdoor activities in unfamiliar areas where stinging insects are known to occur.

2. When outdoors, don't use floral-scented cosmetic products or leave sweet beverages or foods exposed in areas where they might attract bees.

3. Avoid garbage collection areas, which attract hymenoptera.

4. When outdoors, always wear shoes and, if possible, a long-sleeved shirt, long pants, or other protective clothing. Don't wear clothing with a bright floral print or loose-fitting clothing in which stinging insects may be trapped.

5. Don't make rapid movements around stinging insects or intentionally disturb either the insects or their nests.

6. Eliminate all hymenoptera nests around inhabited areas (wild honeybee colonies may be removed by a local beekeeper). For nests that are easily detectible, spray with diazinon, carbaryl, bendiocarb, diazinon or pyrethrum. Spray after the sun has set and first direct the stream at the nest opening and then soak the nest. Commercially available pressurized wasp killer cans are very effective, having almost instantaneous knockdown power. For nests in walls or in the ground, dust with powders of any of the insecticides described above, but with yellow jackets, finding the nest is very difficult. Yellow jackets

may be partially controlled with insect baits made of microencapsulated diazinon mixed with hamburger or tuna fish, but be careful that your pet doesn't get the poisoned bait.
7. Educate young children on the hazards of venomous arthropods.
8. Hypersensitive individuals can achieve some degree of protection from allergic reactions by getting repeated injections of a weakened extract made from whole bee bodies and bee venom. However, such desensitization treatment does not guarantee immunity, and one recent study reported that 40 percent of desensitized people still exhibited allergic responses to bee stings.

THE DIPTERA: FLIES, MOSQUITOES, GNATS AND MIDGES

Of all animals that cause humans pain, disease, suffering and death, none can even come close to the order of insects knows as the Diptera. The Diptera, as their name implies, have only two transparent wings, the front pair. The hind wings have been replaced by small knob-like balancing organs called halteres. There are 17,000 species of Diptera in North America, and they are very widespread, ranging from just below the Artic Circle southward to the Panama Canal. Some are beneficial but others take their toll in blood, painful bites, itchy welts, sleepless nights and disrupted vacations. Some Diptera also transmit debilitating and lethal diseases. The Diptera discussed in this book are all equipped with piercing-sucking mouth parts and are bloodsuckers. They are: the biting flies, gnats, sand flies, midges and mosquitos.

SMALL BITING FLIES

Black Flies

Black flies, also called turkey gnats because they transmit parasites to turkeys, or buffalo gnats because of their humpbacked appearance, belong to the family Simuliidae. Seventy-five species of these tiny (about 3-mm-long) stout-bodied daylight hunters are found in the United States, Canada and Mexico. Although commonly called black flies, their color ranges from black to gray to yellow. They have broad, short wings, and the females have blade-like jaws and piercing stylets with which they can suck blood. These vicious little bloodsuckers stab their bayonet-like mouth

Black Fly

parts into the skin and inject their salivary juices, which may contain a numbing local anesthetic. Thus, when they first bite, all that can be seen is a small drop of blood, but as the local anesthetic wears off, a very intensely painful, extremely itchy lump forms on the skin.

Black flies breed by the millions wherever there is running water. Swarms of adult females attack persistently, even following their victims for several miles. They literally form a cloud as they attack en masse. They concentrate their attack on eyes, ears, nostrils, wrists and knees. There have been reports of extreme panic reactions to the attack of a black fly swarm, and some sensitized individuals have gone into anaphylactic shock after getting repeatedly bitten.

Females lay egg clusters on sticks, rocks or plants below clean running water. The eggs hatch in several days and form aquatic larvae, which later emerge as adults. Black flies are widespread in northern states and Canada, but can also be found in Mexico and Central America in isolated areas near water. They are most numerous in early summer, and their numbers decline during the later summer.

Symptoms: The initial bite is often painless but may leave a small drop of blood. Within an hour, a very painful, itchy wheal forms, which may last several days. Later these bites may form blister-like lumps or hard, pus-filled lumps that may last for weeks or even months. Deaths are rare but have been reported.

In Central America and elsewhere in the world, simulid flies transmit a parasitic worm infection, onchocerciasis, that eventually causes blindness. In Africa, millions of people have been blinded, but Canadian and American simulids do not cause river blindness.

First Aid Treatment: Wash with soap and water, and when itch starts, use anesthetic creams containing benzocaine. Soothing lotions such as calamine may also help. Oral antihistamines may help control the itching. Scratching can produce secondary infection, so if the skin is broken use a topical antiseptic.

Prevention and Control: Space sprays or foggers can provide temporary protection of a camp site, and repellents like "deet" also provide individual protection for several hours. Mechanical protection—gloves, long sleeves, long trousers—prevent bites. Brimmed hats fitted with very fine mesh sprayed with "deet" will keep the little suckers away from your eyes, nostrils and ears.

Biting Midges, Gnats

Tiny mosquito-like flies of the family Chironomidae are nighttime pests that swarm in vast numbers as the sun goes down. They have common names like "punkies," "no-see-ums," "midgets" and gnats. They are tiny (1 to 3 mm in length) and often can penetrate even very fine screening. While they bite viciously, the discomfort produced is much less severe than the bites of black flies. They are found around water, in moist woods of the Atlantic coast and northern and Canadian woodlands, and gather in swarms or clouds at night.

Punkie

The raised, itchy but not particularly painful lumps last several hours, but some victims may show delayed responses of round, elevated, pus-filled blisters that last a few days.

Control: Space foggers afford temporary protection. Repellents do work and protective clothing is a must. Spray repellent on screens.

Sand Flies

The fly family Psychodae, genus *Phlebotomus*, are tiny, hairy blood-sucking gnats commonly known as sand flies. The females of these small (1.5 to 4 mm in length) flies have piercing-sucking mouth parts and are bloodsuckers that feed on many animals, including humans. They are active only at night when there is practically no wind. They are very weak, noiseless fliers and move in short hops of several inches. They seek shelter in dark protected areas and often invade cabins. They usually hide in dark corners and feed only after the sun has set. They can usually get through normal screens and insect nets. Sand flies are widely distributed worldwide wherever breeding places combining darkness, humidity and decaying organic matter are available.

Sand Fly

Their bites are similar to those made by "no-see-ums," itchy raised bumps that itch for several hours, some delayed lumps. In some parts of the world, sand flies transmit a number of diseases such as Carrion's disease, leishmaniasis and sand fly fever. Leishmaniasis has been reported in tropical Mexico and Central America. Northern species don't carry disease.

Prevention and Control: Space sprays are very effective against these very weak fliers, and phlebotimid flies are very sensitive to residual sprays as well. Very fine netting sprayed with repellent is also highly effective, as is personal protection with "deet."

LARGE BITING FLIES

Horseflies, Deerflies and Stable Flies

Large, biting, bloodsucking flies are among the most annoying persistent pests. Their bites are instantly painful as they imbed their bayonet-like stylets into the skin with strong thrusts of their bodies. Stable flies make even larger wounds by scratching with their teeth to encourage blood flow. The narrow feeding tube injects saliva into the wound, and then blood and tissue juices are pumped into the flies' digestive systems. If undisturbed, they can suck up a full load of blood in three to four minutes. However,

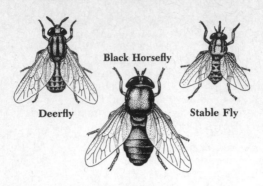

since the bite produces a sharp pain, they usually are brushed away but attack again with remarkable persistence. They are noisy, buzzing, rapid, long-range flies. Only the females suck blood, and they bite only during the day. The stable fly is only as big as its cousin the common housefly (both are Muscidae) but has a broader abdomen and is usually found around barns and other places where domestic animals are found. Another predator on warm-blooded animals is the deerfly. This fly is larger (about 10 mm long), more vicious, more numerous and even more persistent than the stable fly and, like the stable fly, can transmit disease from animals to humans. The largest of the biting flies, the horsefly, is about 20 mm long and attacks both domestic animals and humans. It is noisy, large and slow and is easily swatted into oblivion. There are over 100 species of these nasty, biting Tabanid flies. They are stout-bodied, with large, often banded, colorful eyes; Hence some are called "green heads." Among them is the famous "blue-tail fly" of folk-song fame. Deerflies can be distinguished by large yellow spots on a black abdomen. They become household pests wherever domestic or wild warm-blooded animals are found in the vicinity of the home.

Symptoms: Immediate sharp, highly localized pain at the site of the bite that may leave a tiny red macule where blood leaked into the bite area. This is followed within several minutes by a raised, inflamed red patch that is painful and sometimes itchy for an hour or so. Hours to days later, the bite may form a delayed immune type skin reaction with local swelling.

Diseases transmitted by these flies are numerous. For example, there was a recent outbreak of "deerfly fever" or pneumonic tularemia in Martha's Vineyard, Massachusetts. Other diseases transmitted by these flies include anthrax and loa loa.

Prevention and Control: Space sprays don't do much good, because these are all strong fliers that cover considerable distances. Repellents are effective, particularly those with high "deet" contents. The mouth parts of these flies are powerful enough to pierce single layers of fabric; hence layered clothes afford more protection.

MOSQUITOES

Mosquitoes, dipteran insects of the family Culicidae are long-legged, slender, delicate, humpbacked pests that are widely distributed from just below the Arctic Circle to the Equator. The females have a long piercing proboscis with which they suck blood and leave itchy, swollen wheals. Most of the time they are simply a persistent annoyance, but some also transmit deadly diseases: malaria, dengue fever and yellow fever. There are 34 genera and 2700 species of mosquitoes, but we will concentrate on only three subfamilies that are commonly encountered: the *Anopheles*, the *Culex* and the *Aedes*.

All these mosquitoes lay rafts of eggs in standing water, and these eggs hatch into air-breathing larvae called wrigglers. After several molts, the wrigglers form pupae that attach to the water surface prior to emerging as adults. The adult female usually mates within a few days after emerging and then sets out seeking a blood meal. Her hunting is activated by sensors that respond to the concentration of carbon dioxide in the air. Once airborne, she is attracted to warm, moist objects, and once on the skin, she decides whether to feed. The piercing-sucking organ, the long proboscis, houses a collection of piercing stylets and a sucking tube. Penetration takes from about 30 to 50 seconds, and bloodsucking may last for a bit more than two minutes. Withdrawal of the proboscis takes only five seconds, and a female with an abdomen full of blood flies off and won't feed again for two days.

The largest mosquitoes, the black and white gallinippers, are found in the eastern United States from Canada to Florida and the Gulf Coast. They breed in puddles or water left in marshes at the high spring tides. Like most mosquitoes, they are most active at night and they produce a buzzing song which to most of us is about as pleasant as a fingernail drawn along a blackboard. There are a number of salt marsh mosquitoes of the genus *Aedes*, some of which fly only short distances, a mile or so from their breeding grounds, while others can cover 75 miles in six weeks. Some mosquitoes, such as *Culex quinquefasciatus*, the southern house mosquito, and the Yellow Fever mosquito, *Aedes aegypti*, breed indoors in flower vases, water pitchers or any other standing

Culex Mosquito

Aedes Mosquito

Larva

Anopheles Mosquito

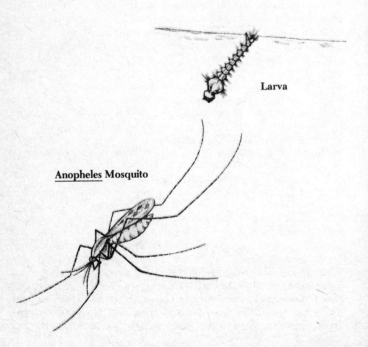

water. Both Culex and Aedes mosquitoes have their bodies parallel to the surface when they feed, but the *Anopheles* mosquitoes hold their proboscises and bodies in a straight line and look like tiny nails being driven in at an angle into the skin.

Symptoms: Mosquitoes inject their saliva when they have a meal, and this always produces a local swelling that is going to itch; you will scratch and sometimes cause a secondary infection. Some people who have been bitten may frequently develop a degree of immunity to local mosquitoes, but may still react to foreign species. Some individuals have very strong reactions and considerable swelling beyond the bite area.

Prevention and Control: Mosquito control starts at the breeding grounds and is really best handled at the community level, but homeowners have some tools for local control in their immediate dwelling and its surroundings. First is to provide good mechanical barriers—that is, well maintained screening (16- to 18-mesh). Secondly, eliminate breeding sites around the house by covering or emptying any water receptacle in the immediate vicinity of your home, e.g., sewer inlets, roof gutters, catch basins, tree stumps and sumps. Ornamental ponds are frequent breeding sites, but you can keep your pond's beauty and still control mosquitoes by introducing a few mosquito fish or guppies that gobble up the larvae. Fill hollow stumps or tree holes and turn empty flower pots upside down. Fogging the area with pesticides containing malathion, fenthion, chlorpyrifos or pyrethrins provides good knockdown of adult mosquitoes. Mists containing carbaryl, malathion, methoxychlor or pyrethrins sprayed into dense shrubbery where the adult mosquitoes hide have also proven effective. Space sprayers or foggers may provide a home site with temporary relief, but best results are obtained with mechanical barriers such as layered clothing and netting or screens. In heavily infested areas, repellents sprayed on clothing and netting help because some mosquitoes can get their proboscides through thin, tight fabrics. Repellents don't stop biting but do prevent landing. They apparently act by blocking the mosquitoes' sensory apparatus. Good repellents containing "deet" will provide a few hours of protection.

FLEAS (SIPHONAPTERA)

Fleas are minute, hard-bodied, eyeless, wingless insects that are flattened bilaterally. They have piercing-sucking mouth parts with which they suck blood. Females require blood to form their eggs. The eggs eventually fall off the host and fall into resting places (sand, flooring, rugs, etc.). The whitish, active larvae feed on

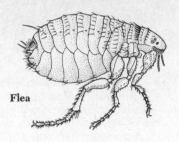

Flea

organic debris and then form cocoons, which eventually develop into adult fleas. The adults can remain quiescent in the cocoons for weeks. Fleas are very susceptible to dryness and thrive best in high humidity. They are quite resistant to cold (about -1° Celsius), and hence can stay over winter in a state of suspended animation.

Fleas have been around in their present form for at least 50 million years, and some 1500 species of fleas are known. They have evolved powerful jumping legs to aid them in reaching their host. The jump of the flea is lightning fast, too fast for the human eye to follow. Fleas weigh less than a thousandth of a gram, are about one or two millimeters long and can jump 100 times their own body length. Human fleas can execute standing jumps of more than one foot. They seem to be tireless and have been reported to jump 600 times per hour for as long as 72 hours. In the jump the flea may somersault or spin while airborne, but most of the time they land on their feet. Cat fleas *(Ctenocephalides felis)*, dog fleas *(C. canis)*, human fleas *(Pulex irritans)* and rat fleas *(Xenopsylla cheopis)* all attack humans, the last mentioned being the transmitter of the dreaded bubonic plague that killed millions of people during the Middle Ages.

Fleas can be encountered across all of North America but thrive best in humid areas.

Control: To control flea infestations, one must hit all parts of the life cycle. The eggs, which usually fall off the host, are small, translucent, shiny, oval structures which can be seen with the naked eye. They hatch in one to twelve days, depending on the humidity and temperature, and form legless, white, maggot-like larvae which, depending on environmental conditions, persist in the larval condition from a week to over half a year. These larvae feed on organic matter, dislike light and need high humidity, so they seek protected moist areas. They then weave cocoons which may last a long time before the adult fleas emerge.

Since the flea's life cycle is so dependent on the environment, the first thing to do is remove potential hot spots such as shady areas under houses where pets go to escape the heat. These should be screened off. Since pets are the usual source of household flea infestation, sanitation of the pets' living area by washing and frequent vacuuming to get rid of eggs, larvae and pupae is indicated. Pay particular attention to upholstery, with all its cracks and crevices. Most pets, including mine, are very sneaky and will only get up on beds and sofas when you are not around, so be sure to vacuum furniture carefully. Baseboards and spaces between floorboards are other favorite spots. After vacuuming, put the vacuum bag outside or in the incinerator.

Recently, a new product, methoprene, produced by Zoecon Corporation, has shown promise. It is an insect growth regulator that disrupts the development of the pupa into an adult flea. While this product will disrupt the life cycle, it won't get rid of the biting adult fleas.

Chemical control of adult fleas by pesticides can be attained by means of a number of commercially available pesticides. Treatment should include outdoor areas that are flea reservoirs frequented by pets. There are aerosol spray preparations and fumigant aerosols which require sealing the room. Remember, whenever doing it yourself, read the label and follow instructions carefully. Dusts, wettable powders and aerosols of diazinon, malathion, propoxur, pyrethrum, bendiocarb, carbaryl and DDVP have all proven to be effective. When applying indoors, be very thorough, particularly when treating rugs; hit the entire surface and lift the edges and apply to the undersurface.

If adult fleas are present, some short-term protection can be afforded by using a repellent on your skin and clothing (see repellents on page 78). Finally, you should prevent reinfestation by making sure your pet is flea-free. Flea collars are important, and flea powders and baths will keep your pet flea-free.

One last word about flea infestations. The family automobile can become infested, and it is very difficult to get rid of fleas once they get in the car. Automobiles provide ideal nooks and crevices for eggs, larvae and pupae. You can spray all surfaces with commercially available residual sprays or close the windows and use an aerosol fumigant. Both of these materials are usually available at pet shops. Repeat treatments may be needed, since flea infestations in automobiles are remarkably persistent.

BEDBUGS

The most important of the bloodsucking true bugs are the bedbugs, which probably first became acquainted with man when he

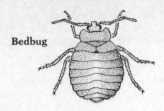

Bedbug

shared caves with bats and swallows during the Ice Age, and which have since become fully domestic.

The two species that attack humans are *Cimex lectularius*, the common bedbug, and *Cimex hemipterus* (formerly *rotundatus*), or the Indian bedbug. *C. lectularius* is found throughout the temperate regions of the world, while *C. hemipterus* is the abundant tropical bedbug. In cold weather, bedbugs hibernate and do not feed, but in warm climates they are active the year around. *C. lectularius* is, however, sensitive to temperatures above 96°F, and if the humidity is high, they die.

Bedbugs are round, flat insects of a rich reddish-brown "mahogany" color, which has led to their being called "mahogany flats." They have short, broad heads with prominent compound eyes, long four-segmented antennae and a three-segmented beak or proboscis. Their legs are well-developed, and they can crawl up vertical surfaces with little difficulty. Bedbugs are almost wingless and are four to five millimeters long when fully grown. One of the most striking characteristics of bedbugs is their peculiar foul, pungent odor produced by "stink glands" located at the bases of the hind legs. Bedbugs have piercing and sucking mouth parts, which are folded back under the head and thorax when not in use and extend downwards at a right angle to the body when the bug is feeding. As the bedbug feeds, it injects saliva, which has the effect of partly digesting the food and so making it more fluid. The saliva also contains an anticoagulant. Bedbugs seldom cling to the skin while sucking, preferring to remain on the bedding or clothing.

Bedbugs are night prowlers, hiding away in cracks and crevices during the daytime. Favorite hiding places are in bedsteads, in the crevices between boards, under wallpaper and similar places for which their flat bodies are eminently adapted. They sometimes go considerable distances to hide in the daytime, and show remarkable resourcefulness in reaching sleepers at night. In fact, most of their activity is directed towards searching for a host. The antennae are held out straight in front of the head, and evidence suggests that they are guided towards a suitable host by following up a temperature gradient.

Symptoms: Although some people seem to be immune to the effects of bedbug bites, others experience considerable irritation. Children suffer most from bedbug bites. In susceptible persons, there is severe irritation caused by the bug's salivary secretion. The bite produces a swollen, itchy red blotch with a central spot. In severe cases, there may be a marked nervous reaction, accompanied by digestive disorders and loss of sleep.

First Aid: Because bedbugs attack at night while the host is asleep it is not really possible to administer immediate first aid. There is nothing that can be done but treat the symptoms by applying calamine lotion, mild analgesics, a paste of baking soda and cold cream or a compress moistened with a diluted solution of ammonia.

Prevention and Control: Prevention of bedbug infestation consists chiefly in good housekeeping, but occasional temporary infestations are likely to occur in almost any inhabited building.

Bedbugs and their kin are susceptible to a variety of pesticides, e.g., 0.5% diazinon, 1-2% malathion, 0.5% DDVP and 0.5% pyrethrins. Seek out their hiding places, such as buttons and seams of mattresses. Spray, but do not soak, and let the mattress dry before replacing the cover. Pyrethrins have little residual effect, hence will require several repeated weekly treatments. Do not use pyrethrins in areas inhabited by people with ragweed allergy. Do not treat infant bedding with pesticides! Use the sprays to hit potential bedbug hiding sites, such as wall crevices, cracks, baseboards, moldings and bedsteads. If the house has a bat or bird bedbug-like infestation, spray cracks around window frames, attics and ceiling light fixtures to prevent more of these pests from migrating into your living quarters.

Evidence of Infestation: Bedbugs are quite democratic in that they will set up housekeeping in the homes of the well-to-do and immaculate or the poor and less careful. They can gain entrance to houses or apartments by stowing away in furniture, bedding, laundry, old books and even salvaged lumber. Because these bugs are so crafty in finding hiding places, their refuges often defy easy detection; however, their glands do emit a distinct, obnoxious sweet odor which may give away their presence. Another way of detecting their presence is by their calling cards, blood stains and spots of excreta (yellowish-black spots of partially digested blood). However, the most immediate warning of the presence of these bloodsucking pests is the presence of bedbug bites.

Related Species of Cimex That Prey on Humans: Bats that get into homes and roost may introduce a number of species of Cimex that can overrun their roost and become a real and annoying household pest when they migrate into human living areas. Birds that build nests on or in human dwelling places may introduce bedbug-like pests into the home, particularly in the fall, when the birds abandon their nests to migrate. The hungry bugs in search for a new warm-blooded food source may lead to heavy infestation of the home. The symptoms of their bites are like those of bedbugs. Controlling these pests comes down to bat-proofing your home and removing birds' nests attached directly to the house.

LICE

Lice are minute, grayish-tan, bloodsucking, wingless, dorso-ventrally-flattened insects whose size and shape allow them to move through densely packed hairs. They have effective piercing-sucking mouth parts which are retracted into the head when not in use. The ends of their legs are equipped with a curved claw with which these pests cling to hairs of their host. Lice fasten their eggs to hairs with a glue-like substance. These tiny attached eggs are called nits; hence, the term "nit-picker." There are three Anoplurida that parasitize humans: the head louse *(Pediculus humanis capitis)*, the body louse *(Pediculus humanis corporis)* and the crab louse *(P. pubis)*.

Several important human diseases have been transmitted from person to person by such lice. These include typhus, relapsing fever and trench fever. However, most human infestations just produce intense itching. While head lice afflict all kinds of people, females are more frequently affected, and children have a greater risk of exposure than adults. American blacks seem to be less

Crab Louse

Human Body Louse

Hog Louse

frequently affected. Infestation by lice can result from contact with bedding, toilet seats, using someone else's comb and from sexual intercourse. Clothing is a good vehicle for transmission to a new host, since eggs are sometimes laid in fabric hence the term originating in World War I of "seam squirrels" to designate lice.

Symptoms: Infestation with lice, sometimes called pediculosis, produces intense itching and, in very severe cases of chronic infestation, skin damage and anemia. Sometimes the scalp becomes encrusted, due to secondary infestation.

Treatment of Lice on the Body: One percent Lindane lotions and shampoos have both proven to be effective. The shampoo should be left on for five to ten minutes and the lotion for about half an hour. These are available by prescription only and should not be re-applied for seven days. Other lotions that are effective against body lice are 10% crotamiton or pyrethrin-containing lotions. The lotion should be put on for five to ten minutes and then washed off. People who are known to be allergic to ragweed should not use pyrethrins. Malathion lotions (0.5%) left on the scalp for twelve or more hours are also effective. If the eyebrows and lashes are infested, the above-mentioned treatments may be too harsh for the delicate eye tissues. See your physician, who will prescribe petroleum jelly (which suffocates lice) or physostigmine (eserine) ointment.

Prevention and Control: Spray suspected areas with sprays containing malathion (2%). Lice can be detected by using an ultraviolet light source, since they fluoresce (glow white) under this light and their movements can be seen.

ASSASSIN BUGS

There are more than four thousand species of Reduviidae, a family that usually feeds on insects but in some cases feeds on humans. One close relative of these insect assassins, the Triatomas, not only feed on human blood but spread a parasitic disease (Chagas disease) in Central and South America. Like all true bugs, the reduviids live on a fluid diet (plant juices or animal blood). To obtain this food, these sucking insects use a beak or proboscis that pushes a set of sharp, hollow, hypodermic-like piercing stylets through the skin. The bug then injects a venomous saliva that contains enzymes that break down fats and tissues, thus liquifying them so that they can be sucked in. In animals, the main effect of the poison is to release histamine, a potent vasodilator. The

Assassin Bug

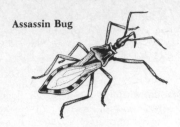

venom is also a potent antigen, and because of this the bite can produce extremely severe allergic reactions in victims previously exposed to bites by these insects. There are several reduviids common to North America that have been reported as household pests that attack humans. These are conenose bugs and wheel bugs.

CONENOSE OR KISSING BUGS

Species of the genus *Triatoma* have the elongate (cone-shaped) head which is characteristic of the family Reduviidae. Hence, the name "conenose bugs" is often used to describe these insects. They range in color from light brown to black and may exhibit checkerboard-like orange and black markings where the abdomen extends laterally past the folded wings. These insects are flattened, which allows them to hide in small cracks and crevices. One conenose, *Triatoma sanguinosa*, also called the Texas Bedbug, is found from Maryland southward and westward to Kansas. These bugs are usually associated with raccoons and oppossums that take up residence in crawl spaces or attics of homes. Another conenose, *T. protracta*, is an occasional pest in Texas, New Mexico, Utah, Colorado and California. This pest is usually associated with wood rat infestations. The size of mature conenose adults varies from approximately one to three cm (0.4 to 1.2 inches) in length, depending upon the species. Most give off an intense, foul, sweet odor.

Conenose bugs are nocturnal insects. They take their blood meals at night and hide in any available crack or crevice between feedings. Assassin bugs, as a group, normally feed on small mammals, but in the absence of their preferred hosts, several species will readily feed on humans. The proboscis contains four piercing stylets that easily penetrate the skin without producing any initial pain. They are commonly referred to as "kissing bugs" because in 1899, a woman in Washington, D.C. was bitten on the lip by a *Triatoma*, which led to wide newspaper coverage and a nationwide hysteria about "kissing bugs." Some of the common sites of human attack, in order of frequency, are the hands, arms, feet, head and trunk. If hosts are available, these bugs feed every three or four days, but these conenoses can survive for several months without feeding. In warmer climates they remain active throughout the year. While the bite itself is painless, the reduviid, when startled while feeding, will quickly withdraw its mouth parts, causing a severe, painful wound.

Symptoms: Triatoma bites produce a spectrum of allergic reactions that are dependent on previous history of bites by these pests and subsequent sensitization. The lesions—and there may be a cluster of them—are more severe than more common insect bites and consist of swollen, red, pimple-like eruptions with a central dot of blood. In sensitized individuals, giant urticarial lesions with a central puncture occur as a generalized reaction. Sometimes a delayed reaction several days after the bite produces hemorrhagic nodular to bulbous lesions of the hand and foot.

Prevention and Control: Remove animal nests from around the house (see page 12). Otherwise follow same procedures as those listed for bedbugs.

WHEEL BUGS

The wheel bug *(Arilus cristatus)* has the typical small, narrow head characteristic of the Reduviidae family. They are mouse gray in color and are approximately from 1 to 1.6 inches long. A coglike crest on the dorsal side of the prothorax is distinctive to this insect and accounts for its popular name, "wheel bug." They are generally found in the southern two-thirds of the United States.

Wheel bugs are usually predators on soft-bodied insects. Human bites are usually the result of accidental contact while handling vegetation, boards or other objects. The bug penetrates the skin with its "beak" or proboscis and injects a potent venomous secretion normally used in killing its insect prey. Human bites are characterized by immediate intense pain that usually subsides in from three to six hours. Treat such wounds with topical corticosteroids. Antihistamines will relieve inflammation. Analgesics like aspirin or Tylenol may help relieve the pain.

Control: The best way to prevent wheel bug contact is to be able to identify this unusual insect and avoid it. Children should be instructed not to handle it. Wearing leather gloves while working outside will prevent bites that occur when the wheel bug is accidentally picked up with vegetation or other debris. Since wheel bugs are predacious on many harmful insects and are generally considered beneficial, control is not recommended.

INSECT REPELLENTS

Early in the 1930's, it was discovered that the chemical ethyl hexanediol was effective in repelling a number of biting-sucking insects and arachnids. This substance is still in use in the commercially available repellent 6-12. An even more potent repellent, N, N, diethyl-meta toluamide, was discovered in the 1950's, and this substance, commonly called "deet," is used in most present-day repellents. It is effective against chiggers, ticks, fleas, mosquitoes and biting-sucking flies, but affords no protection from stinging ants, wasps, bees or hornets. According to the U.S. Army, the Department of Agriculture and the Environmental Protection Agency, "deet" is both effective and non-toxic except in some rare individuals with certain genetic metabolic diseases. It can, however, sting when it gets in the eyes or mouth or onto cuts or rashes. Repellents may be applied as a spray (either hand pump or aerosol), as a solution, as a stick or as a cream. Towelettes are also available, but they are wasteful since each towelette can cover only a small area.

According to consumer reports (June 1982) not all repellents contain equal amounts of "deet," hence vary in effectiveness. Muskol has the most, 95%, followed by Repel (52%), Cutters (31%), Deepwoods Off (29%), Off (19%) and 6-12 (10%, but it also has ethyl hexanediol). The most economical and easily carried forms of repellent are the liquids and creams. Follow instructions carefully and when in heavily infested areas cover exposed skin areas and spray clothing, since many insects can bite right through fabric.

CENTIPEDES

We have included centipedes in the section on biting and stinging animals because they are equipped with poison glands and fangs. Indeed, there is one large (up-to-six-inches-long) centipede, *Scolopendra,* found in the southwestern states, that can inflict a painful but not serious sting; however, most centipedes have such weak jaws that they can't even break human skin.

Centipedes belong to the phylum Arthropoda, class Chilopoda. They are nocturnal, have flattened, elongated, segmented bodies with one pair of legs per segment. At their head end, there is a

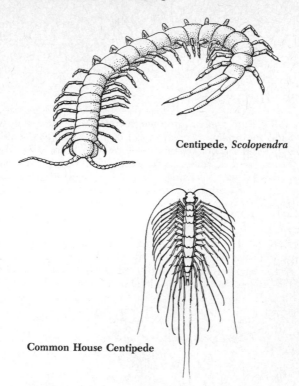

Centipede, *Scolopendra*

Common House Centipede

pair of long. slender, many-jointed antennae. The common household centipede, *Scutigera coleoptera,* is about an inch long, yellow-brown, striped and wormlike in appearance; it is capable of fleet, darting movements with its fifteen pairs of long, slender, multijointed legs. In females, the last pair of legs are longer than the rest and project backwards.

Centipedes in the home are really quite beneficial, as they prey on insects and spiders and probably are valuable allies in pest control. Except to their six- and eight-legged victims, they really do no harm, but most thinking bipeds view them with revulsion, repugnance and fear. In fact, you may not want them sharing your living quarters.

Control: The simplest thing to do to control centipedes is to eliminate their breeding places, which are usually moist litter and refuse in basements and around windows, window-wells, doorways and woodpiles. Bendiocarb wettable powder made into a spray mixture and spritzed onto these places should keep centipedes out. If you want to get rid of centipedes in the house, spray their usual hiding areas—e.g., baseboards and piles of firewood—with a residual pesticide such as propoxur.

BEETLES *(Order Coleoptera)*

Beetles are sheath-winged insects that occupy a great variety of habitats and comprise the largest order of insects; approximately 40 percent of all insects are beetles. Only two types of beetles directly harm humans, but many indirectly affect us as destructive pests on plants, lawns, foodstuffs and wood. The North American species that are harmful because they secrete a blister-forming (vesicating) agent belong to the family Meloridae.

Blister Beetles (family Meloridae)

The Meloridae are narrow, elongate beetles characterized by a "neck" (pronotum) which is distinctly narrower than its head or wings. Adult beetles range in body length from one to two cm (0.4 to 0.8 inches) and vary considerably in their coloration.

Blister Beetle

Distribution: In general, blister beetles are found in greater numbers in the eastern half of the United States.

Habitat and Behavior: The immature stages of the blister beetle usually feed on other insects and are not harmful to man, but adult blister beetles release a clear amber fluid by rupture of thin membranes in the leg joints or other segmented areas of the body. This fluid contains a vesicating (blister-causing) agent called cantharidin, and is triggered by pressure against the body of the beetle. Light pressure exerted by clothing or by brushing off a beetle is usually sufficient to cause the release of its vesicating fluid. The adult blister beetles are readily attracted to bright white light, and many cases of human exposure occur at night around such lights. Since the adults are plant feeders, some cases of human vesication occur when handling flowers or vegetation infested with blister beetles.

Control: In areas with an abundance of blister beetles, use yellow light bulbs for outdoor lighting. If a meloid beetle lands on the skin, blow it off; do not crush it. Since cantharidin is distributed throughout the beetle's body, crushing the beetle against exposed skin would result in maximum cantharidin exposure. Skin irritation resulting from blister beetle contact is seasonal, with the greatest number of vesicating incidents in the United States occurring in July, August and September.

The reaction generally consists of a superficial line of blisters on the skin. These blisters do not require emergency treatment.

Bombardier Beetles (Brachinus spp.)

Are capable of forcefully ejecting a spray of boiling hot, extremely irritating quinones from glands at its rear end. The release of this defensive spray is explosive, and the beetle produces it by secreting two harmless substances—peroxide and quinone—into the mixing chamber, where there is an immediate, heat-releasing explosive reaction which releases oxygen, thus forcing the boiling irritant spray out with a distinctly audible explosion. In Kenya, bombardier beetles cause a symptomology called "Nairobi eye" when they shoot this spray into the face of a hapless victim. These beetles can accurately aim their spray and often respond to accidental brushing against a leaf that they are on.

Giant Water Bugs *(Benacus)*

One of the hemipterans that has adapted to a freshwater environment is the giant water bug *(Benacus)*. These bugs have not developed gills and frequently come to the surface for air. These immense predators are also strong flyers and can fly from one water body to another. They are attracted to light and are sometimes called "electric light bugs." They are so powerful that they have been known to kill frogs and will even attack birds encountered in flight. Their offensive weapon is a very strong retractable beak which can inflict a very painful wound. These insects can reach lengths of up to ten cm, are brownish and may be encountered in the water (they are powerful swimmers) or around lighted campsites, tennis courts and baseball diamonds.

These bugs are found around many bodies of fresh water from

Giant Water Bug

southern Canada to Panama, particularly near freshwater ponds in the vicinity of well-lighted porches and homes. Human bites are relatively rare.

Symptoms: An immediate sharp pain, minor inflammation.

First Aid: Treat as you would any insect sting.

ARACHNIDS: SPIDERS, SCORPIONS, TICKS, CHIGGERS AND MITES

Few creatures are as universally abhorred and unusual as spiders and their kin. They evolved over 300 million years ago and are indeed remarkable. They have an external skeleton and eight legs with 56 leg joints, and can perform miracles of locomotion, readily moving backwards, forwards, up, down and sideways. Most are equipped with poisonous fangs—chelicerae—with which to immobilize their prey. Foods are predigested outside of the body and then sucked into the spider's bowel. Spiders may have as many as a thousand spinning tubes with which they weave webs of varying dimension, complexity, and sometimes beauty. This webbing is extraordinarily light, yet incredibly strong.

Of the 50,000 to 70,000 known species of spiders in North America, only a few are harmful to humans, with only three poisonous spiders posing an actual threat. The chief culprit among them is *Latrodectus*, the infamous "Black Widow," a shiny, black, sedentary, shy creature that only bites in self defense. Although as many as 2000 black widow bites may occur in a year, they account for only a few deaths.

Almost everybody fears spiders. They often occupy our bad dreams. Like Little Miss Muffet, many people have arachniphobia (fear of spiders). Brave men have been sent screaming by a creature they outweigh by some million times. The medieval mass hysteria over fear of the bite of the fuzzy tarantula gave rise to music and frenzied dance, the tarantella. Actually, most tarantula bites are no worse than bee stings, although there is one Mexican variety of tarantula that is considered mildly harmful.

While most American scorpions are relatively harmless, some found in the southwestern United States, some Caribbean islands and Mexico are truly to be feared. A few decades ago, the notorious Durango scorpion of Mexico was responsible for over 1200 deaths per year.

Finally, the real scourge of mankind have been the lesser lights among the arachnids. Ticks, chiggers and mites have not only caused us itchy days and sleepless nights but have also been the vectors for a number of deadly diseases. Scrub typhus and Rocky Mountain spotted fever are two of the worst ones.

BED MITES

An unseen and uninvited occupant of your mattresses and pillows is an abundant, tiny mite that feeds on human body dander, dead skin cells, that we continuously shed. These denizens of our mattresses are virtually indestructible and are included in the section with harmful Arthropod house pests because they cause many people a variety of allergies that range from runny eyes, stuffy noses and itching skin to serious asthmatic conditions. This creepy critter is *Dermatophagoides*, a cosmopolitan inhabitant of beds of the rich and poor alike worldwide. It was not until the 1920's that physicians discovered that house dust was a potent allergen that was highly specific. Because of seasonal variations in its presence, investigators began to look for some biological process, and in 1964 a Dutch team of researchers discovered *Dermatophagoides*, the bed mite, and proved they were the source of the allergen in house dust.

These mites are microscopic, eight-legged, spine-chilling (to see) creatures with unsegmented, flattened bodies bearing hairy feelers. Their thin, outer skeleton allows these mites to extract water from the air, and humidity sensitivity is the weakness in the mite's armor. These mites mate only once or twice in a lifetime and produce 20 to 40 eggs, which hatch in about six days; the hatchlings reach maturity in 25 days. Given this prodigious reproductive capacity, there are literally hordes of these little bugs in the top layer of your mattress. They do quite well when the humidity is high—50 percent or more—but in the dryness of northern households in the winter their populations crash and the number of mites decreases dramatically. Come the spring rains and the increase in humidity, the survivors don't take long to repopulate your bed. The allergen, a single polypeptide, is found mainly in the mite's feces (80%) and the rest in their dead carcasses. This material is so light it becomes airborne as microscopic particles, and when inhaled by sensitized individuals produces the classic spectrum of allergic responses.

Prevention and Control: It is literally impossible to get rid of all of the bed mites, since they are virtually indestructible. It may be possible to reduce their numbers by lightly spraying the upper surface of your mattress with lindane and pirimiphos-methyl; however, allergic individuals might develop sensitivity to the pesticide. Best results are obtained by frequent vacuuming and keeping the humidity low. See your allergist, since it may be possible to desensitize people to house dust.

ITCH MITES

Itch mites are Arachnoidea, belonging to the family Sarcoptidae, and are also called sarcoptic itch mites or scabies mites. These whitish skin parasites have a round body and no distinct head, but do have jaws. They have eight legs, the two front pairs equipped with suckers and the two rear pairs with long bristles. They are almost invisible to the naked eye, being only 0.4 of a mm long. Females are larger than males and are responsible for scabies itch. The mite is usually acquired by personal contact, but can also come from shared living quarters. The female, once on the skin, attaches itself with its suckers, raises on its hind end and eats its way into the skin in 30 to 40 minutes. Once in the skin, the female burrows out a tunnel, where she deposits eggs. This tunneling goes on for about sixty days, at speeds of 1.4 to

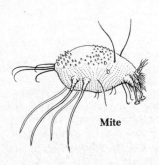

Mite

two inches per day, and then she dies. The eggs hatch; the larvae leave the tunnel and move over the skin.

The tunneling in its initial stages does not produce itching; but after a month or so, the toxic secretions of the female cause inflammation and intense itching. This itch has been called "the Norwegian itch" or "the seven-year itch." Apparently the mite is highly selective about where she burrows, preferring (63%) webs between fingers and where the wrist flexes, (2%) the armpit), (11%) the elbows, (9%) the feet and ankles, (8%) the penis, buttocks and nipple area in adult females.

Diagnosis is based on distribution, symptoms and detection of the tunnels, which appear as tiny gray or brownish lines in the skin. The secondary stages involve red hives, lumps and sometimes scabbed-over wounds.

The mites are very widely distributed worldwide, but can survive for only three days to two weeks in moist, cool air. Their best habitat is human skin. Thus, wherever there are people there can be itch mites.

Prevention and Control: A number of other mites that attack humans have been introduced by their primary hosts: rats, bats and birds. As long as these vertebrate pests inhabit your home, these mites are content to feed on their usual host, but once they are gone, the mites seek out new victims: you and your family. The best way to control these pests is to prevent vertebrate pests from setting up housekeeping in and on your home.

TICKS

Ticks belong to the same class of animals as spiders, scorpions and mites, the class Arachnoidea. They are small, external, bloodsucking parasites found all over the world. They have few natural enemies and a prodigious reproductive capacity; hence they can be very numerous. As adults, they have eight legs that stick out prominently from a flattened, leathery, egg-shaped body. They are capable of becoming quite distended when filled with their host's blood, particularly the females, which are considerably larger than the males. The mouth consists of two small, retractile jaws (mandibles), two short appendages called palps, and a central probe equipped with recurved teeth (the hypostome). These structures are attached to a plate called the capitulum. Ticks attach to the host with their mouth parts, which are not only imbedded in the skin but are also glued in place by a cement-like secretion. Sometimes, when pulling off an imbedded tick, the capitulum

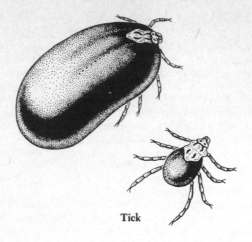

Tick

and its attached parts remain in the skin and cause a local reaction. Ticks can voluntarily detach from the host.

Ticks have sense organs that enable them to detect mammals as much as 25 feet away. They respond to shadows, touch and odors, and either drop onto or, on contact, latch onto a passing victim. Once on you, they climb to a comfortable spot, often the head, and imbed their mouth parts and start to suck blood. During this time they inject their saliva, which contains a variety of proteins including anticoagulants and a toxic substance. There are three species of ticks that commonly attack humans: the Lone Star tick, the American dog tick and the wood tick. They are most active between March and August, when they become sexually active. Females will remain attached to their host for from five to 13 days during this period and then detach themselves.

Ticks are found over much of the United States and Mexico from sea level to high woodlands and meadows below the frost line. They are found on stalks of grasses, weeds and shrubs as well as trees.

Symptoms: Ticks can transmit a number of diseases caused by microorganisms, such as rickettsia. They cause Rocky Mountain spotted fever, Lyme's disease and Q fever. Their toxin can also cause tick paralysis or tick toxicosis. Tick paralysis has occurred in British Columbia, the northwest and the southeast United States. It is not common and occurs mostly in children, where it is a potentially fatal disease. The toxin is a nerve poison. Children, most usually female, complain of tingling in the legs and have trouble walking. They become irritable and weak and lose their appetites. Twenty-four to 48 hours later, there is a rapid progression of sensory disturbances, followed by paralysis, which can be pronounced in the tongue and face. The paralysis may be so severe that respiratory paralysis and death result.

Treatment: If symptoms described above occur, make a careful search for ticks and remove them by painting with kerosene, nail polish remover or gasoline. The tick will usually let go.

Control: Keep weeds and grass around the home well cut to remove sites where pets can pick up ticks. Provide your pets with fresh tick collars and, during the tick season, inspect your pets for ticks. Once a gravid female gets off your pet and produces eggs, you have a problem in the home. To prevent this, dust pets with 5% carbaryl or malathion powder, particularly around the head, neck, ears and toes. If young ticks are in the home, spray cracks, crevices, baseboards, furniture, rugs (including the undersurfaces) and draperies with 0.25% bendiocarb, 1% malathion, 2% ronnel or 0.5% diazinon. Be particularly attentive to areas where your pet rests. Tick infestations are tough to treat and repeats may be needed. Fumigation is not suggested for control of ticks in the home.

CHIGGER MITES

Chiggers, also knows as "red bugs" or "jiggers," are one of the few of 35,000 species of mites that prey on humans. Like all members of the arachnids, they have eight legs and a pair of piercing jaws or chelicerae. They have a complicated life cycle, and it is only in the larval stage that they attack humans. In cooler climates, the larvae are present only in the summer months, but in southern, warmer climates (60°F or warmer) they may thrive through most of the year. These larvae are almost at the limit of

Chigger

human vision, being 1/50th of an inch long when not fed. Chiggers live in dense population islands in grassy fields, meadows and scrubby savannahs. Often one can sit or stand on such an island and be heavily attacked, while another person only a few feet away will not encounter a single chigger. The chigger crawls onto its victim and seeks out a favorite spot, often where clothing is tight or where parts of the body are in contact, e.g., the groin, belt line, ankle, back of the knee or armpit. It chews into the skin with its chelicerae and injects its saliva, which digests human skin cells, which are then sucked up. As the chigger bores deeper and becomes progressively engorged, a tube called a stylosome is formed. It is the stylosome tube and the body's reaction to it that cause the awful itch.

The chigger remains on its victim for three to four days before dropping off of its own will. The itching starts within hours and is worse at night. In the early stages of attack, all you see is an itchy bump with a red pinpoint, the chigger, in its middle. Within a day it forms an intensely itchy hive with a tiny, fluid-filled, pinhead-size blister in its center. Even after the chigger departs, the lesions and itch persist intermittently for seven to ten days. The chigger sores may be almost an inch in diameter and look like chicken pox, only they are highly localized.

Prevention: Treat both clothing and skin with repellent such as "deet" (see page 69). Chiggers may remain on clothing for some time; hence field clothing should be washed in water heated to 100°F.

THE BLACK WIDOW SPIDER *(Latdrodectus Mactabs)*

One of the two dangerous species of spiders in the United States and southern Canada is the shiny, black, female black widow. This spider is not aggressive but will bite if mildly provoked or when guarding a nest with egg sac. The venomous female is identifiable by the well-known red hourglass or red dots on the underside of the abdomen. The brownish male is small, one-third the size of the female, and, although venomous, has such weak fangs that it cannot pierce human skin. Although black widows are sometimes found in the home, most often they are found in dark, moist brush piles, wood stacks, sheds, barns and outhouses (often under the hole). The last-mentioned locus is an important one to bear in mind, since about 50 percent of all bites are on the genitals and are acquired while sitting in privies.

Black widows, sometimes in considerable numbers, are found

Black Widow Spider

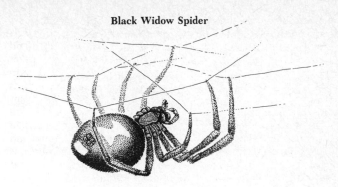

in all states except Alaska, and in southern Canada, Mexico and Central America. They are most prevalent in warmer climates but can also be present in cooler environs and are found from sea level to up to 8000 ft. Most bites occur in June, July and August. Females hibernate in winter, and bites are rare during that time. Most fatalities (and they are very few in number) occur in the southeastern United States.

Symptoms: The poison of the female black widow is a neurotoxin. Drop for drop, it is 15 times more potent than rattlesnake venom. However, the amount injected is usually quite small; hence there is little mortality.

At first the victim will feel a sharp sensation not unlike a pin-prick. The wound is not obvious, with little swelling or tissue destruction, and requires careful search—use of a magnifying glass is helpful. The area bitten may display two tiny red holes with slight local swelling. Within 15 minutes, symptoms appear. A numbing or dull pain slowly develops and increases in intensity, reaching a maximum of excruciating pain in from one to three hours. The pain may continue for 12 hours and/or up to two or more days before subsiding just as slowly. The poison acts on the nerve endings, affecting the thighs, chest, shoulders, back and in particular the abdomen, which may become rigid and boardlike. Other possible reactions are sweating, nausea, respiratory distress, salivation, weakness, vomiting and convulsions. Hematocrit, blood and cerebrospinal fluid pressures are often elevated. Symptoms usually subside within 48 hours. Only 5 percent of black widow bites are lethal. Best estimates (of several years ago) indicate that no more than half a dozen deaths per year might be attributed to this spider. Today, with improved diagnosis and treatment, there are few, if any, deaths in the United States.

Healthy men or women usually survive, but children, the elderly and those with high blood pressure are at high risk.

First Aid: Do not use a tourniquet, incision or suction, as the poison spreads too rapidly for these measures to have any effect. Placing ice packs on the wound may reduce pain but other first aid is useless. The victim, if under 16 or over 65, must see a physician as quickly as possible.

Treatment: Immediately call your poison center for instructions.

Control: Frequent cleaning to remove spiders and their webs from buildings and outdoor living areas will decrease the possibility of accidental bites. Routine hosings down of potential spider habitats, such as under steps and around windows and doors, will discourage the black widow from locating in these places. When working in spider-infested areas, wear gloves and a long-sleeved shirt. If warranted, the outside or inside of a building may be sprayed with an approved insecticide, particularly organophosphates and carbamates in the form of dusts or sprays like bendiocarb, diazinon, malathion and pyrethrins. Apply the spray around windows and stairs, in closets or other spider habitats such as cracks and crevices, in accordance with instructions on the manufacturer's label. Retreatment 30 days later will kill newly hatched spiderlings.

BROWN RECLUSE SPIDERS *(Violin Spiders)*

The brown recluse, *Loxosceles reclusa*, and its related species have become the most widespread and dangerous spider inhabitants of houses in the United States. These shy, reclusive arachnids have taken up residence in closets, attics, furniture, woodpiles, stacks of old papers and even inside boots.

The brown recluses are relatively small spiders that as adults range in size from a quarter of an inch to three/eighths of an inch in length. They are usually a dirty yellow-brown or light brown in color, and their frontal section, the cephalothorax, is smaller than the rear section, the abdomen. They have six eyes, in pairs, and eight slender legs, each terminating in two small claws. All brown recluses bear a violin-shaped, dark brown to black spot on the upper surface of their frontal section; hence they are commonly called fiddlebacks or violin spiders. Their bodies are slightly hairy. The female is about half again as big as the male and is the usual biter.

The female constructs messy-looking, irregular webs in out-of-the-way places and feeds on all varieties of insects, which she rapidly dispatches with her potent venom. She lays eggs from May through August, producing thirty to 300 spiderlings in a breeding season. The life span of these spiders usually ranges from about 300 to 900 days. They are hit-and-run specialists that bite if inadvertently disturbed and then quickly retreat. Most bites are on the hands and arms. The initial impact of the bite is barely noticeable—perhaps a mild stinging—and there may be two puncture marks where the fangs have penetrated the skin. However, the spider's toxin is potent, and the aftermath of the bite is where the real problem lies. From two to eight hours later, the bite site becomes reddened and painful, sometimes severely so. A blister develops, surrounded by a ring of inflamed tissue that sometimes consists of alternating rings of whiteness (blanching) and redness, giving the lesion the appearance of a target. Within three to four

Brown Recluse Spider

days, the blister becomes a firm swelling, up to ¾ of an inch in diameter; this then develops into an ulcer. In adult victims there could be accompanying fever and chills. As is always the case, older people and young children are more susceptible and may show severe symptoms such as joint pain, hives, nausea, vomiting, hemolytic anemia and thrombocytopenia. Blood may appear in the urine, and in some cases kidney failure may occur. Even if the symptoms subside, the ulcer formed at the bite site tends to heal partially and then erupt again and again. These bites are potentially serious, and you should see a physician who should in turn call the local poison control center for advice.

Control: See black widow.

TARANTULAS

There are several genera and species of large, hairy, fierce-looking spiders that are commonly referred to as tarantulas. These relatively harmless but king-sized spiders have been used by the makers of horror movies to exploit people's general terror of spiders—a terror that is hardly deserved, because in most cases the bite of a tarantula is not more serious than a bee sting. These large, sluggish spiders are usually quite docile and are sold in many pet stores to people with unusual tastes.

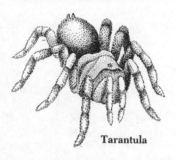

Tarantula

There are some small, native, hairy spiders such as *Atypus snetsingeri;* although among the smaller of the tarantulas, they are impressive, having half-inch-long bodies and a fearsome appearance. In the early summer months, the male native tarantulas in their search for mates may wander into your home, but, while they can bite, their venom is mild and no cause of particular concern.

OTHER HOUSE SPIDERS

The most common biting spider in the United States is a small, hairy, very aggressive, short-legged, crab-like jumping spider of the genus *Phidippus*. There are several species, some of which have white stripes and fluorescent green mouth parts. Sometimes these are mistaken for baby tarantulas, but, unlike tarantulas, these are quite pugnacious and bite with little provocation. They also tend to hang on, thus rendering them susceptible to a good swat. Their bite is sharp and painful and will raise a pale swelling surrounded by red within minutes and produce a dull, throbbing pain. As the swelling subsides, there can be considerable itching at the bite site. The symptoms will go away spontaneously in a few days, but if you are anxious that the spider might have been a black widow or a brown recluse, then pickle the dead spider in rubbing alcohol, gin or vodka and bring it to your local university or agricultural service for identification.

All spiders are capable of biting, and all contain some form of venom; however, in most cases their bites are of little consequence. The spider most often found in your home will be the cosmopolitan house spider, *Achaearanea tepidariorum*, which builds webs everywhere, particularly around basement windows. The main problem for the homeowner is not the bite of the house spider but rather its dust-catching, abandoned webs that clutter up corners of our ceilings and windows.

Control: Spiders are good friends to have in your home, since they prey on many insect pests. However, if you find them repugnant, you can use the same control methods described for black widows on page 90.

SCORPIONS

There are about 6000 species of scorpions found throughout the temperate tropical regions of the world, and, despite what you've seen in the movies, most of these are not dangerous. Scorpions, like spiders, ticks and mites, are arachnids. They have eight legs, with the front pair held forward and bearing pincers like those of

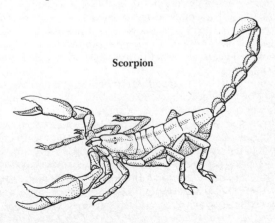

Scorpion

Scorpion

a crab, used to grasp insect prey, while the other six legs are used for walking. The tail or abdominal portion of scorpions is composed of five segments, the hindmost of which is bulbous and contains a poison gland and a very sharp, prominent, hypodermic-like stinger. The common striped scorpion, which will frequently invade homes, does sting if touched, but the sting is no worse than that of a bee, producing local inflammation, pain and swelling.

The common striped scorpion is about two inches long, the males being slightly larger than the females. It is identifiable by two thick dark bands that run the length of the back. The markings are the same for both sexes. These nocturnal hunters are found through most of the southern United States, preferring dark hiding places such as crevices and cracks around window- and door-wells, under debris, woodpiles, crumbling brick and stone walls and under the bark of dead trees and logs. They enter houses through cracks or gaps in windows and doors and are frequently brought in with firewood. The female produces about thirty live young, who ride on her back for two weeks or so before they take off on their own.

In limited parts of Arizona there dwells one potentially deadly scorpion, *Centroides sculpturatus*, a narrow, two-inch-long, semi-transparent, yellow-brown killer whose venom is a rather potent neurotoxin that can kill up to 25 percent of children who are stung and untreated. Not too many years ago, *Centroides* scorpions were responsible for 1500 to 2000 deaths a year in Mexico, but an education campaign, bounties and the destruction of breeding sites has reduced the scorpion threat there. If you are stung by one of these Arizonian *Centroides*, immediately call the poison control center for advice.

Control: Start by sealing crevices and cracks in foundations, since these are the usual entrance sites. Clean up potential scorpion breeding places around the house, e.g., old boards, woody debris, etc. Treat areas where scorpions might gain access to the house with sprays or paints of a residual pesticide, such as propoxur 70% wettable powder (two ounces per gallon of water), Diazinon 4E (0.5% spray) or bendiocarb (one packet/gallon).

PART III

Insect Pests That Attack Wood

Wood has been one of mankind's most valuable materials. It has been used as a readily available fuel and has provided us with a remarkably strong and durable construction material. Under proper conditions it may last thousands of years, and in other conditions can be destroyed rapidly. There are only a few organisms that can digest wood: bacteria, some fungi and some insects. These organisms have developed enzymes that specifically attack cellulose, the main constituent of wood. Wood is made up of 50 percent cellulose, which is very similar to and yet very different from starch. Both starch and cellulose are composed of simple sugar molecules linked together, but the nature of the linking is different. Most animals can digest starch, while the sugar-to-sugar linkage in cellulose resists digestion. Those organisms that have the enzymes to break down cellulose and other wood constituents into simple sugars are capable of doing enormous damage to structural timbers and furniture. Fortunately, most insects that attack wood in the forests don't do much harm to wood that is well dried and used for furniture or construction, but in wood that has a high moisture content or is infected with a fungus, the damage can be enormous. The principal villains, of course, are termites, but there also are some beetles that can do extensive damage. These will be covered in detail in the following pages.

In addition to insects that do damage by eating wood, there are several insects that bore into wood in the process of building their nests there. The most important of these is the large, prolific and highly destructive carpenter ant. Finally there are a number of insect pests that infest firewood and are often brought into the home with bark-covered logs. Each of these pests will also be dealt with separately in this section.

The following section on subterranean termites is based on U.S. Dept. of Agriculture Home and Garden Bulletin No. 64, by H. R. Johnston, Virgil Smith and Raymond Beal.

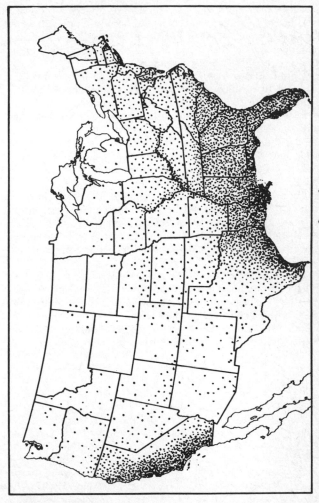

Relative hazard of termite infestation in United States is indicated by density of stippling.

SUBTERRANEAN TERMITES: THEIR PREVENTION AND CONTROL

Subterranean or ground-nesting termites are by far the most destructive insect pests of wood. They attack buildings and other wood products in all states of the Union except Alaska, but are most common and aggressive, and hence most destructive, in the temperate parts of the country (see map).

The United States now has about 35 million single dwelling units subject to termite attack, and only a small percentage of these units are treated annually to control termites.

The total cost of termite control is estimated at not less than $750 million per year. One-third of this cost is believed to be for damage repairs; the other two-thirds is for chemical treatment. The total does not, however, represent the entire impact of termites on wood in use. In many instances termite damage goes undetected, and in numerous others the homeowners either will not or cannot do anything to stop the damage. The losses in these categories probably far exceed the amount expended for control. Furthermore, termites do considerable damage to utility poles, fence posts, and other wood used for similar purposes. The annual costs resulting from termite damage and its control probably exceeds $2 billion.

Rising costs are expected in the future due to four factors. One is the expected population increase—more houses mean more opportunities for economic losses to termites. The second is the discovery of the Formosan termite in a number of Gulf and Atlantic port cities. This species is far more aggressive than our native species and may eventually extend its range over much of

Termite

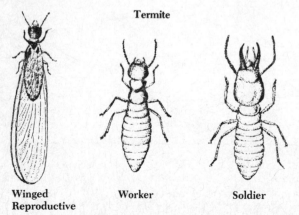

Winged   Worker   Soldier
Reproductive

the United States. The third is the annual rate of inflation, and the fourth is the ban of species-effective pesticides that have potential toxic impact on humans.

Buildings can be safeguarded against termite damage through efforts in the planning stage and during construction. Preventive measures in all new construction and effective control measures wherever termite infestations develop will decrease the waste of wood and wood products. Also, these measures will save the homeowner much anxiety and expense later.

Increased use of concrete and masonry terraces adjacent to foundation walls and poorly designed slab-on-ground construction favor termite attack and result in mounting damage to buildings.

After a building has become infested with termites, it is often difficult and costly to apply effective control measures. Difficulties and costs vary among buildings. An infested building should be carefully examined to determine the extent of the infestation and the measures needed to prevent further damage. Some infested buildings require only simple structural changes, repairs, or chemical treatments, all of which can be made by the owner. Others may require such major changes or complicated treatment that he may need the services of a specialist who knows the habits of termites and is experienced in their control.

BIOLOGY OF SUBTERRANEAN TERMITES

Termites are primitive social insects of the order Isoptera that are descendants of a cockroach-like ancestor of the Coal Age. It is from these roach-like ancestors that termites derived their ability to digest cellulose with the aid of microscopic protozoan animals and bacteria that they house in their hindguts. These cooperative protozoa produce enzymes that split cellulose into simple sugar molecules, which are then used by both the protozoa and the termites. In short, the termites provide dwelling place and food for the protozoa, and the protozoa pay rent to their hosts in sweet currency.

There are more than 2000 species of termites in the world, mostly in the tropics. In the United States, there are two types that cause damage: subterranean termites and dry wood termites. This section is devoted to the more destructive of the two, the subterranean termite.

Subterranean termites are social insects that live in nests, or colonies, in the ground. Each colony is made up of three forms or castes—reproductives, workers, and soldiers (see illustration). During their lifetimes, the individuals of each caste pass through three stages—egg, nymph, and adult. The adult workers and soldiers are wingless, grayish white, and similar in appearance. The soldiers, however, have much larger heads and longer mandibles, or jaws, than the workers. The worker is an insect that destroys wood and is the one usually seen when a piece of infested wood is examined. The soldiers guard the colony. The reproductives, or sexual adults, have yellow-brown to black bodies and two pairs of long, whitish, translucent wings of equal size. (They differ from the reproductive forms of ants, which have two pairs of transparent wings of unequal size.) These wings are shed after the mating flight. Termites have thick waistlines, while ants, which are often mistaken for termites, have thin waistlines.

DISTRIBUTION

Subterranean termites are found practically throughout the tropical and temperate parts of the world. They are common throughout most of the eastern half of the United States and along the Pacific coast. The termite hazard varies greatly within a general area. In any specific locality, it depends upon such factors as nature of the soil, moisture conditions and local building practices.

Subterranean termites probably have existed in their present distribution for millions of years. There is no evidence of their general introduction or spread from the tropics to the United States, or of widespread movement of any of our native species from the southern to the northern States. Infestations of buildings, however, have become more common with the general adoption of central heating plants, since heated basements can prolong the period of termite activity. This fact, together with other changes in building practices and use of construction materials, explains why termites have become a problem in areas where formerly they were of little importance. The vast development of suburban homes in forested areas likewise has aggravated the termite problem.

DETECTION OF TERMITE INFESTATION

Large numbers of winged reproductive termites emerging or swarming from the soil or wood may be the first indication of the presence of a termite colony. Even when the actual flight of these adults is not observed, the presence of their discarded wings is very good evidence of a well-established colony nearby. These discarded wings often are found on the floor beneath doors or windows where termites have emerged within a building and have been unable to escape.

Termite damage to wood often is not noticeable on the surface; the exterior surface must be stripped away in order to see the extent of damage. The workers avoid exposure to air by constructing galleries within the materials which they attack. Occasionally, they completely honeycomb wooden timbers, leaving little more than a thin shell. The inside of their galleries is covered with grayish specks of excrement and earth. Subterranean termites do not reduce the wood to a powdery mass or push wood particles to the outside, as do certain other types of wood-boring insects (see page 111).

The presence of flattened, earthen shelter tubes—arcades—which these insects construct over the surface of foundation walls is another sign of termite infestation. These tubes are from one-fourth to one-half inch or more wide. Termites use them as passageways between the wood and the soil from which they obtain essential moisture. The tubes also protect termites from the drying effect of direct exposure to air.

DEVELOPMENT OF A TERMITE COLONY

Flights of termites occur most frequently after the first warm days of spring, often following a warm rain. They may also occur at almost any time during the spring or summer, and sometimes even in the fall, especially in warm parts of the country. In buildings with heated basements, termites occasionally fly during the winter. The individuals in these flights are adult winged reproductives, sometimes called kings and queens, that have developed in well-established colonies. They are attracted by strong light, and when they emerge within buildings they gather about windows or doors. Here they soon shed their wings. Then, pairs of termites try to return to the soil to find a suitable location for starting a new colony.

Most of these perish, but some pairs survive and succeed in hollowing out small cells in or near wood in the ground. After this is done, the female begins laying eggs. During the first year she lays only a few. The young termites hatching from these eggs are cared for by the parents and develop into workers and soldiers. These gradually take over most of the duties formerly performed by the original royal pair.

Egg laying increases rapidly in a termite colony after the first two or three years. Secondary reproductive forms, without wings, also develop and lay eggs, which serve to supplement those of the original queen. A colony more than five or six years old may contain the royal pair, secondary reproductives, soldiers, and thousands of workers.

CONDITIONS THAT FAVOR TERMITE INFESTATION

Subterranean termites become most numerous in moist, warm soil containing an abundant supply of food in the form of wood or other cellulose containing material. They often find such conditions beneath buildings where the space below the first floor is poorly ventilated and where scraps of lumber, form boards, grade stakes, stumps or roots are left in the soil. Most termite infestations in buildings occur because wood touches or is close to the ground, particularly at porches, steps, or terraces. Cracks or voids in foundations and concrete floors make it easy for termites to reach wood that does not actually touch the soil. Termite activity is increased and prolonged, even in northern areas, when soil within or adjacent to heated basements is kept warm throughout most of the year.

Conditions under which termite colonies thrive are rather rigid. More is required than the mere existence of a wooden structure. Much depends on how the structure is built. For example, unless it is possible for the termites to maintain contact with the ground or some other source of moisture, they will die. Because of this, certain steps taken during the construction of a building will greatly reduce or prevent future termite damage.

KINDS OF MATERIALS DAMAGED BY TERMITES

The principal food of subterranean termites is cellulose, obtained from wood and other plant tissues. Termites, therefore, feed on wooden portions of buildings, utility poles, fence posts or any other wood product. They also damage paper, fiberboard and various types of fabrics derived from cotton and other plants. Many noncellulose materials, including plastics, may be penetrated and damaged by termites, even though they do not serve as food. Termites also occasionally injure living plants. The greatest economic loss, however, is to the woodwork of buildings.

PREVENTION OF TERMITE ATTACK DURING CONSTRUCTION

The best time to provide for protection against subterranean termites is during the planning and construction of a building. This has been learned through research on the habits and behavior of termites and through experience in their control.

Improper design and construction of buildings, resulting perhaps from lack of knowledge or of indifference to the termite problem, are favorable to infestation. It is therefore important that every effort be made to stress the value of good building practices and chemical soil treatment during construction. These practices are described in detail in the U.S. Department of Agriculture Home and Garden Bulletin No. 64 which is available from the U.S. Government Printing Office.

CONTROL OF TERMITES IN EXISTING BUILDINGS

Ridding existing structures of termite infestations and making them resistant to future infestation is a major problem in termite control. Generally, buildings become infested because little or no attention was paid during their construction to preventive measures that would have made them resistant to termites. It is in such buildings that termites cause heavy losses each year.

PERIODIC INSPECTIONS

Make periodic inspections of buildings for evidence of termite attack. This is simply good insurance and should be done regardless of how completely preventive measures were employed in construction. The frequency of such inspections will depend on the abundance of termites in the area and the type of construction. In areas of extreme hazard, inspections should be made annually.

SANITATION PRACTICES

Sanitation and structural control measures should be given first consideration in the control of existing infestations. These control measures include the following:

1. Remove all wood, including form boards and other debris containing cellulose, from underneath and adjacent to buildings with crawl spaces.
2. Remove other wooden units, such as trellises, that connect the ground with the woodwork on the exterior of the building. Replace them in such a manner as to break these contacts. Impregnate wood piers and posts that will be driven into the soil, using an approved preservative applied by a standard pressure process.
3. Replace heavily damaged (structurally weakened) sills, joists, flooring, etc. with sound material. Where possible, remove all soil within 18 inches of floor joists and 12 inches of girders.

4. Fill voids, cracks or expansion joints in concrete or masonry with either cement grout, roofing-grade coal-tar pitch or rubberoid bituminous sealers.
5. Provide adequate drainage and ventilation.

CHEMICAL SOIL TREATMENT FOR EXISTING HOMES

Formulations of four pesticides—aldrin, dieldrin, chlordane, and heptachlor—are currently registered for use in treating soils to control native subterranean termites. In south Mississippi tests, these chemicals, applied at the prescribed rates and methods, have provided complete protection for 24 to 28 years. To date, no alternative materials have been found that will provide comparable long-term, economical protection.

Preparation of Chemicals

A soil chemical is economical and most easily prepared when purchased in the form of a *liquid concentrated solution*. The concentrate is sold according to the percentage, or weight in pounds per gallon, of the toxicant it contains. These percentages and weights vary according to the amount of toxicant present in the concentrates of the different chemicals. Each concentrate contains an emulsifier to make it mixable with water, and *must be diluted before it is ready to use*.

Directions for diluting the concentrated solutions to the strength of the finished emulsion recommended are usually given on the container. In the event that they are not, the following directions should be used in preparing each chemical for soil treatment.

1. *Aldrin, 0.5% in water emulsion.* Aldrin is usually sold as a liquid concentrate containing either two or four pounds of the technical grade chemical per gallon. To prepare a 0.5% water emulsion, ready for use, dilute one gallon of the two-pound concentrate with 47 gallons of water, or one gallon of the four-pound material with 95 gallons of water. This makes 48 gallons of the 0.5% water emulsion from the lower concentrate and 96 gallons from the higher one. The rate of dilution is one to 47 and one to 95, respectively, regardless of the unit of measure used (gallon, pint, etc.).
2. *Chlordane, 1.0% in water emulsion.** Chlordane is sold as 46–48- or 72–74-percent liquid concentrate. To prepare a one-

*In some states, chlordane is not available to the public and only licensed pest control operators have access to this potentially toxic chemical. Check with local officials to find out what pesticides are allowed and available.

percent water emulsion, ready for use, dilute one gallon of the 46% concentrate with 48 gallons of water, or one gallon of the 72% material with 99 gallons of water.

3. *Dieldrin, 0.5% in water emulsion.* Dieldrin is usually sold as a liquid concentrate containing 1.5 pounds per gallon of the technical grade chemical. To prepare a 0.5% water emulsion, ready for use, dilute one gallon of the concentrate with 36 gallons of water.

4. *Heptachlor, 0.5% in water emulsion.* Heptachlor is sold as a liquid concentrate containing two or three pounds of the actual chemical per gallon. To prepare a 0.5% water emulsion, ready for use, dilute one gallon of the two-pound concentrate with 48 gallons of water or one gallon of the three-pound concentrate with 72 gallons of water.

Some of the main ways to control termites by chemical treatment in existing buildings are as follows:

1. Slab-on-ground houses.

Termite infestations in houses built with a slab on the ground present serious control problems. It is difficult to place chemicals in the soil beneath such floors where they will be effective. One way to do this is to drill holes about half an inch in diameter through the concrete slab close to the points where the termites are, or where they may be entering. Space the holes about six inches away from the wall and approximately 12 inches apart to insure proper treatment of the soil underneath. *Take care to avoid drilling into plumbing and electric conduits.* Apply the chemical through the holes by any practical means available. Another way is to drill through exterior foundation walls to the soil just underneath the slab and introduce the chemical through the holes. This method is complicated, however, and usually requires special treatment by a trained pest control operator.

2. Raised porches, terraces, and entrance slabs.

Termite infestations occur frequently in porches, terraces, and entrance platforms. The most satisfactory way to control infestations at these places is to tunnel under the concrete slab *adjacent to the foundation wall,* all the way from one side to the other, and apply a chemical in the bottom of the tunnel, or trench. Remove all wood debris encountered in digging the tunnel. Place an access panel over the opening to permit annual inspections and additional soil treatments, if needed. Another way to treat this area is to drill holes 12 inches apart, either through the adjacent foundation wall *from within the crawl space or basement,* or through the entrance slab, and introduce the chemical through these holes.

3. Crawl-space houses.

Buildings with crawl spaces usually can be treated easily and effectively. In general, the following procedures can be used:

  a. Dig trenches six to eight inches wide adjacent to and around all piers and pipes, and along both the inside and outside of all foundation walls. For poured concrete foundations, the trench need be only three to four inches deep. For brick and hollow block masonry foundations, it should be at least 12 inches deep. Where the footing is more than 12 inches deep, make crowbar, pipe or rod holes about one foot apart and extend them from the bottom of the trench to the footing. This will prevent termites from gaining hidden entry to the building through voids in these types of foundations. *The trench should never be dug below the top of the footing.*
  b. Pour one of the chemicals listed on pages 104 and 105 into the trench at the rate of four gallons per ten linear feet *for each foot of depth.* If the trench is a deep one, apply the chemical to alternate layers of about six inches of soil.

4. Basement houses.

To treat the soil along the outside walls of basements, dig a trench six to eight inches wide and about a foot deep, *adjacent to the wall.* Then make crowbar, pipe or rod holes about one foot apart that extend *from the bottom of the trench to the footing.* Pour the chemical into the trench at the rate of four gallons per ten linear feet *for each foot of depth from grade to footing,* alternately replacing and treating six-inch layers of soil.

5. Houses with wells.

In houses where wells are located close to or within foundation walls, the same principles of termite control apply as are recommended for their prevention. The main difference is that in existing buildings, wall voids can be treated directly with chemicals. Reinfestation of treated walls in basements can be prevented by first removing the earth along the outside of the wall from the finished grade to the footing. Then fill the mortar joints with dense mortar and waterproof the outer surface. Fill the mortar joints on the interior of the wall and seal the expansion joints where the wall and floor meet.

DRY WOOD TERMITES

There are several species of dry wood termites, mostly of the genus *Kalotermes,* that may cause extensive structural damage to wooden parts of buildings—and even furniture. As their name implies, they attack dry wood and are particularly important pests in the warmer southern parts of the United States, particularly in the Southwest and California. However, isolated infestations of dry wood termites have been reported as far north as Canada. While they can establish themselves in buildings, they can't live outdoors in colder climates.

On bright, warm, sunny days (temperatures 80°F or higher), swarming occurs. In each area, this appearance of brownish, red-headed, half-inch-long, thick-waisted, winged sexual forms is almost clock-like. The mating flight over, the king and queen shed their wings and find a convenient crack or crevice in some wooden structure. They then excavate a tunnel, which they enlarge into a royal brood chamber; the tunnel is sealed off with a plug made of partly digested wood and feces. Discarded wings and small, elongated, brownish droppings underneath cracks and crevices in window frames, door jambs, rafters or even wooden furniture are evidence of an infestation.

The royal pair and the white, soft-bodied, blind larvae or nymphs that make up the majority of the colony gouge passages through the infested wood, literally riddling it. These nymphs pass through seven developmental stages, eventually forming grotesque, blind, amber-headed, large-jawed soldiers. As a new mating season approaches, the nymphs burrow to the surface, making exit holes up to ⅛ of an inch in diameter. These holes are guarded by large-headed soldiers who block the opening against potential predators—usually ants. These holes or any cracks that may expose the colony are always plugged by the secretions and feces of a nymph. Colonies can exist for years; while they seldom number more than 2000 individuals, their long duration permits extensive damage. The main clue to an infestation is piles of fecal pellets that the nymphs have pushed out of openings of their tunnels.

One dry wood termite, the furniture termite *(Cryptotermes),* attacks floors, woodwork and furniture. These insects are much smaller than other dry wood termites and their colonies are rather small. Their tiny droppings, which are constantly pushed out of their tunnels, are like powdery sawdust; thus these termites are also referred to as powder-post termites. These pests not only infest wood but have been known to tunnel into books and papers. Powder-post beetles (see page 112) also produce a similar powdery sawdust.

Control: Dry wood termites are best controlled by fumigation, which will require the services of a licensed pest control operator.

CARPENTER ANTS

The largest and most destructive of house-frequenting ants are the carpenter ants, all belonging to the genus *Camponotus*. While these pests do not eat wood as termites do, they are capable of removing considerable portions of your wooden architecture as they excavate an elaborate maze of tunnels in which they set up housekeeping. In total structural damage to wood they rate second, exceeded only by the cellulose-ravaging termites. There are

Carpenter Ants

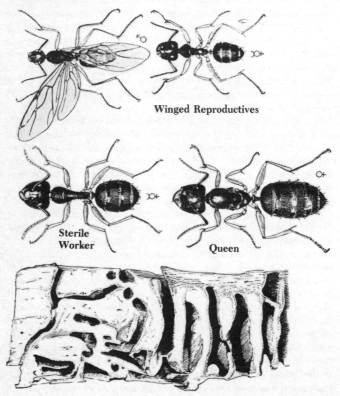

Winged Reproductives

Sterile Worker

Queen

Carpenteer Ant Tunnels

several different species of carpenter ants, and they are widely distributed throughout the United States. The long-legged, swift-moving carpenter ants are dark brown to black in color. Wingless workers range in size from one-quarter inch to one-half inch in length, and females may be up to three-quarters of an inch in length. They are predatory, strong-jawed insects that suck juices from other insects such as aphids. They occasionally invade homes in search of food, but the real problem comes when they decide to excavate ornate, smooth-walled galleries in the timbers of your home.

Large winged males and females emerge from existing colonies from late spring to early summer to engage in their mating flights. The fertilized female or queen finds a suitable nesting site, a hole in wood or a burrow under a stone, and begins to lay eggs that develop into small workers, which then go out to forage for food. They feed the queen, which continues to lay eggs. She is a veritable egg factory. These small workers feed the queen, care for the pupae and work with enormous vigor to build galleries, often in moist wood softened or damaged by decay or in natural flaws in the wood.

Some people with good hearing can actually hear faint rasping sounds from timber infested with carpenter ants in their active tunneling phase. As the workers polish the galleries, they carry away the woody debris and push it out through slits they cut to the outside. The best evidence of their infestation is a rain of fine sawdust-like material. Colonies increase in size and may contain several thousand individuals, including winged males, winged females, the queen and both large and small workers.

Frequently a colony is encountered when splitting wood or when logs with an inconspicuous hole are put in the fireplace and the heat reactivates the hibernating ants, which march out to a fiery demise. These ants can tolerate extremely cold temperatures, and as the weather gets colder they literally produce their own antifreeze.

These ants are remarkably ingenious in the ways they find to invade buildings—climbing in on branches that make contact with the house or even walking on wires. When it warms up, ants nesting near a home often invade it in search of food, but more often the presence of numbers of these large, blackish pest patrolling your rooms is an indication that they have set up housekeeping in your woodwork. The appearance of winged forms in the house is really a warning of infestation of the house timbers, indicating that the rest of the colony are holed up in the wood somewhere.

Prevention and Control: To prevent entry, trim away all branches in contact with the house and survey the exterior for favorite entry places, i.e., where there is wet wood or decay in the wood. Check woodpiles, stumps and damaged trees near the house. These are often the site of carpenter ant colonies, and nests may be found by following the scurrying streams of foraging workers. Be careful of logs brought into the home, being *particularly careful* of bark-covered firewood or logs with holes in the ends. Treat all nests found anywhere near the house to decrease the probability of infestation of the house. Streams of ants entering a cavity in a doorframe or beam indicate that there is a nest inside. Damaged or rotted wood should be replaced, since such wood is an open invitation to carpenter ants. Pay attention to crawl-spaces under the house and check out attics, since carpenter ants may enter under the eaves.

Indoors, baits may reduce the number of foragers, but for effective control the nest or nests must be eliminated. Spray the nest openings with either residual dusts or sprays, using a narrow nozzle to get into the opening. Diazinon and Baygon have proven effective and are relatively safe. Chlordane and heptachlor work very well, but may no longer be available, since there is ongoing controversy about their safety. Pesticides may be introduced into infested wood by drilling holes, but make sure to plug the holes afterwards. Be careful then you do this, since angry carpenter ants swarming out of a disturbed nest can bite rather vigorously.

CARPENTER BEES

Several species of big, handsome, iridescent, deep blue, green or purple, hairless bees of the genus *Xyloncopa* burrow into dry wood. These so-called carpenter bees, depending on the species, range in size from one-half inch to four-fifths of an inch in length; while they are solitary bees, they tend to attack wood in groups, gouging out numerous holes. The males have no stingers, and while the females possess a potent sting, they don't use it very often. However, they do buzz around your head when you disturb their nests. Mating occurs in the Spring; then the fertilized female chews a nesting burrow into wood. She burrows about an inch a week forming burrows that may be more than a foot long. As she digs out her nesting gallery with her jaws, and pushes the resulting sawdust out of the round half-an-inch-in-diameter hole that is the entrance to her tunnel.

The fertilized female brings in a mixture of regurgitated pollen and nectar, lays her egg and seals the cell, after which she starts another. The eggs develop into larvae, then pupae, and by late summer develop into adult bees. As cold weather comes, the adult carpenter bees return to the old nest tunnels to spend the winter.

Control: Carbaryl (5%), bendiocarb (1%) or diazinon (2%) powders blown into carpenter bee entrance holes will effectively eliminate these pests. Later, plug the holes.

WOOD-BORING BEETLES

Relatively few organisms can digest the constituents of wood. Unhappily, two organisms that can perform this feat, termites and the larvae of some wood-boring beetles, do considerable damage to wooden parts of homes and furniture. While beetles are hardly a major problem for the homeowner, they can, if numerous enough, become pests of some significance. Beetles are insects of the order Coleoptera, the order containing the greatest number of species, and there are many different species that have adapted to wood boring. For the purposes of this small book, only those species that are commonly encountered in the home will be considered.

Anobid Beetles

There are several common pest beetles belonging to the family Anobiidae. The "furniture beetle" *(Anobium panctatum)* is the one most frequently encountered. Small (a fifth to a quarter of an inch in length), cylindrical, red-brown to brown adults emerge from infested wood in the warm summer months. As they emerge, they leave circular exit holes less than a tenth of an inch in diameter and push out wood dust as they depart. Thus, holes, dust and the presence of beetles are all good diagnostic signs that you have uninvited guests. Shortly after emergence, they mate. Then

Powder Post Beetle	Anobid Beetles	Old House Border

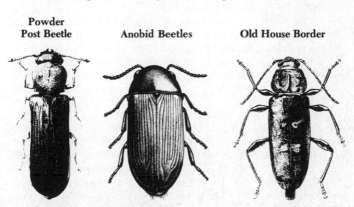

the female starts to deposit eggs on rough, unpainted wooden surfaces. Three to four weeks later, the eggs hatch and the tiny, white, grub-like larva starts boring into the wood with its powerful black jaws. For the next two or three years, it tunnels continuously and grows to about a quarter of an inch in length, leaving behind small balls of droppings. These pests require high humidity; hence in dry, well-heated homes they are hardly a problem except in damp basements or in attics that are a little leaky. They are often introduced into homes in newly acquired antique, unpainted bargains that are stored in basements or attics awaiting refinishing by some enthusiastic but procrastinating collector. Left untended, several generations will spawn, and a serious infestation can develop. Interestingly, the exit holes in wooden *objets d'art* or antiques are considered proof of old age (some antique counterfeiters have been known to use shotguns and blast fine birdshot into a newly manufactured "antique," putting in a pedigree with the perforations).

The deathwatch beetle *(Xestobium rufovillosum)* is another anobid pest of wood—mainly oak. It gets its name from its plaintive love call advertising its availability. In the spring, after the pupae metamorphose into adults, the sexually mature beetles move through their tunnels near the surface of the wood and repeatedly rear up on their legs, banging their heads on the tunnel ceiling. This volley of about ten taps is answered by nearby deathwatch beetles and, in the quiet of the night, this tap-tap-tapping was considered to be a foreboding of impending doom. Perhaps these beetles inspired Edgar Allan Poe. Once the quarter-inch-long, brown beetles emerge, they leave their one-sixteenth-of-an-inch-in-diameter holes and mate, despite the fact that both probably have terrible headaches from all that amorous prenuptial head-banging. The female lives only a couple of weeks, and in that time lays 50 to 70 eggs that hatch into larvae that bore into wood. The larvae tunnel, munch and grow over the next three years or so and may reach a length of almost half an inch. They do well only in moist hardwoods; hence, in modern, dry homes, are problems only in damp basements or poorly ventilated, moist attics.

Powder Post Beetles

There are a number of beetles belonging to the family Lyctidae that are commonly called powder post beetles because of the very fine, powdery dust that they push out of their tunnel openings. The adults of this family are slender, brownish and from a tenth to a third of an inch long. They are most often associated with sapwood of oak, ash, walnut and many tropical woods such as bamboo. A number of tribal masks I brought back from Africa were infested with these pests, which reduced one carving to a mere shell in the course of a year (the average time it takes for the larva to develop). When you find tiny round holes (1/25 of an inch in diameter) and a fine, talcum-like dust, you've got one of the Lyctidae in your wood.

House Borers

House borers are relatively large (up to three inches in length), handsome beetles frequently equipped with very long antennae, hence are referred to as "longhorn beetles." All belong to the family Cerambycidae, and they produce very large (up to four inches long), plump, flat, pale, legless, broad-headed larvae equipped with large, dark jaws. They are frequently introduced into the house with firewood. There are several different species with varying tastes. Some like only old hardwood, while others feed exclusively on new, moist timbers.

The "old house" borer beetle is an attic-dwelling villain that does considerable damage to pine, fir and spruce roof timbers. The dark brown/black, long-horned adults, which bear white patches on their backs, emerge from their tunnels in late summer. The exit holes are large (a quarter of an inch in diameter) and oval. The presence of holes and adult beetles in older homes is fair warning. The larvae are very long-lived—four years or more— and during this protracted period can do considerable damage to structural timbers of a house. If you have keen hearing, you can actually make out the raspy chipping of the larva as it tunnels and, if you happen to possess a stethoscope, you can tune in on some really impressive eating sounds. Not only do the larvae bore into old wood, but the adults, when they bore out, easily penetrate most things in their way, including aluminum siding, ceramic tile and all sorts of shingles.

The "new house" borer is a narrow, black, inch-and-a-quarter-long pest of un-aged pine timbers. The large, long, cream-colored larva tunnels into new wood for about two years and, while the amount of structural damage is minimal, the round, quarter-inch-in-diameter exit holes that appear when the adults emerge are most upsetting to the new homeowner. These borers were already in the lumber prior to construction; hence holes will usually appear only in the first year and then the problem is over.

Control: Anobid and Lyctus beetles can be controlled by painting or spraying the infested wood with 5% pentachlorophenol in an oil base or 1% Lindane in an oil base. Both are usually available commercially—but don't get these solutions on the skin or inhale the vapors. Use a respirator mask, rubber gloves and protective goggles when applying. Also, be careful with the solvents, since they and their vapors are flammable. For small wooden objects such as carvings: wet them, wrap them in aluminum foil and put them in a 200°F oven for two hours. The steam should kill all the larvae and pupae.

Carefully inspect all new wood objects you purchase and keep them under observation for emergence holes and powdery dust. While new house borers are no problem, old house borers can do considerable damage and may require fumigation by a licensed pest control organization.

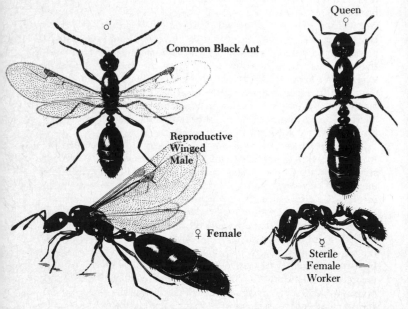

Common Black Ant

Queen
♀

Reproductive
Winged
Male

♀ Female

Sterile
Female
Worker

(The Winged Phase)

PART IV
Kitchen and Pantry Pests

ANTS

Ants are everywhere, and at any given moment there are at least a trillion or so ants living on our planet. This makes them the most abundant of the social insects. There may be more than 10,000 different species of ants, with the greatest variety being in the tropics. These extraordinary ground-dwelling creatures have few natural enemies and they have considerable impact on humans. One species, the carpenter ant (see page 108) can wreak tremendous damage to wooden parts of our homes. Two species can inflict painful stings (see fire ants, page 56, and harvester ants, page 58) and several species have adapted to our homes, where they are mainly nuisances attracted to our pantries and kitchens, where they forage for sugars, starches, grains and fats. While they do relatively little damage, they can introduce bacterial contamination to our foods and are annoying. Ants can also carry certain types of aphids that afflict our attractive and sometimes rather loved and expensive ornamental plants.

Ants are not all bad, and they do prey on a variety of vermin, including bedbugs and other pests. Ground-nesting ants also bring subsoil to the surface, and in the process aerate and moisturize our lawns, although some species can really raise hob with a carefully manicured lawn.

Ants belong to the same order of insects as wasps and bees. Anatomically, they have a well-defined head, bearing large antennae, eyes and chewing mouth parts. Their mid region, the thorax, has six legs and during mating season sprouts wings in both males and females. The hind portion, connected by a "wasp waist," consists of two segments—a narrow pedicel (one or two segments) and a bulbous, swollen portion which in some species carries a potent stinger. During mating seasons, large swarms of winged, recently matured male and female ants are airborne in mating flights. After mating, the expendable males die and the fertilized females break off their wings, find a cavity, seal themselves in (for weeks or even months) and lay their eggs. The eggs are fed by the new queen and develop into small workers who sally forth to collect food, which is brought back to the nest and fed to the queen and the new brood of developing ants that will become large workers. Thus the queen produces successions of eggs, and the workers, which are wingless, sterile females, care for the new broods and nest. Many different outdoor nesting ants can invade the home, but several are particular household pests: the little red ant or Pharaoh ant; the little black house ant or cornfield ant; the odorous house ant; the pavement ant and the thief ant. These will be covered in some detail because of their common occurrence.

The Argentine Ant

The so-called Argentine ant, *Iridomyrmex humilis,* is one of the most persistent and crafty of ants that get into the home. It probably arrived in the United States in the late 1800's, making its entry on coffee shipped to New Orleans from Brazil—hence it should more properly be called the Brazilian ant, but Argentine ant it is. This pest is now not only well established throughout the South and in California but is also popping up in many other places, including Hawaii.

These ants usually nest outdoors in the ground, forming a honeycomb of tunnels and excavating mounds of earth near the exit holes. They are particularly fond of building their nests under wood, and dislodging a plank overlaying an Argentine ant nest will expose enormous numbers of small, 1/10-inch-long light brown workers with yellowish jaws, young pupae and sometimes numerous 1/5-inch-long brown queens. Sometimes they may es-

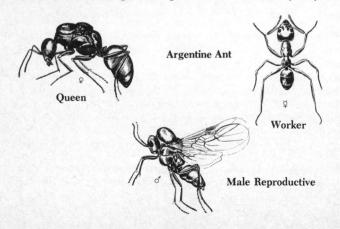

Argentine Ant

Queen

Worker

Male Reproductive

tablish a nest directly in a house, but usually invade houses from outdoor nests—particularly during the warmer months when the rains come. These tenacious, tough pests have no natural enemies that they can't overcome, and they will eat a wide variety of foods, although they show a strong preference for sweets. They farm aphids, scale insects and mealybugs to extract their honeydew, and in the process can do damage to houseplants. They also go after plants that have sugar-secreting nectaries. They forage tirelessly and get through any gap in a covered container.

Control: Control of Argentine ants requires barrier sprays to prevent entrance into the home, removal of plants that attract them (fig trees, oleander, bamboo) and extensive use of sweet, poisoned baits in the fall.

The Pharaoh Ant *(Monomorium pharaonis)*

Pharaoh ants derive their name from the idea that these insects were one of the Ten Plagues visited upon Egypt by God when the Pharaoh would not let the Hebrews leave. This small red to yellow pest was introduced into England from North Africa in the early nineteenth century but quickly made its way across the Atlantic and is now widespread throughout the United States and Canada. Its omnivorous worker scouts range far and wide seeking food sources. When they find one, they leave an odor trail, return to the nest and recruit hordes of other workers to come get the goodies. In a short time, you will see streams of workers traveling to and from the food source. These ants love heated premises and will nest in almost any secluded spot, e.g., between walls, under baseboards or in garden walks near the kitchen door. Their favorite trails are along heating pipes, since they thrive best at about 80°F and in relatively high humidity. There are sometimes several queens in a colony, and new colonies are readily established, leading to large-scale infestations of hundreds of thousands of ants.

Little Black and Red Ants (Genus *Monomorium*)

Two ground-dwelling small ants that nest in lawns, under rocks and walls of buildings and masonry are the little black ants *(M. minimum)* and a little red cousin *(M. destructor)* now widely distributed and frequent invaders of homes during the warm summer months. The red species has been known to bite and is a general nuisance. Both belong to the same genus as the Pharaoh ant.

Thief Ants

Thief ants, tiny cousins of the fire ants, are the smallest of the house-infesting ants. They get their name from the fact that they enter nests of other ants and kill their young; thus they are in part beneficial. The adult thief ant worker is only a millimeter or so long, hence is hardly noticed. They are widely distributed throughout the United States and look like miniature yellow Pharaoh ants. They frequently infest kitchens, pantries and cupboards, where they are persistent pests. They love fatty foods, meats and cheese, etc. but are not big on sweets. Indeed, they can often be found on poisoned dead rats and mice that have given up the ghost between your walls. Their nests, usually in crevices in cupboards, may house several queens; thus their numbers can be considerable. They may also nest outside the home and come in to forage; hence barrier spraying may be needed to keep them out. They are most active during the summer.

The Pavement Ant *(Tenarmorium caespitum)*

These hairy, pale-legged, small, dark-brown ants have furrowed heads and are up to a tenth of an inch long. They are particularly pesky during the summer, when they frequently invade homes in search of sweets. As their name implies, their nests are typically found along seams and cracks in pavements, often forming little sandy mounds with a central crater.

The Odorous House Ant *(Tapinoma sessile)*

When the rains come, the odorous house ant, which is widely distributed throughout much of the United States and Canada, becomes a common and somewhat smelly invader of our homes. It's not that these pests are repelled by water, but that their natural food, honeydew, is washed away by the rains, and so they come indoors seeking new sources of sweets. The tireless workers, about a tenth of an inch long, are brown to black in color and if crushed give off a bad smell. They nest in walls, under walks and under floorboards. These pests can be active throughout the year in well heated homes and are active even at 50°F.

Cornfield Ants (Genus *Lasius*)

Several species of cornfield ants are found throughout the United States and can be serious house pests. They are among the most abundant ants. These ants normally are outdoors species that farm root-feeding aphids for their sweet secretions. The ants carry aphids, burrow into the soil and place the aphids on the roots. Commonly found in and around houses under pavement, in the masonry and even in walls, they frequently invade kitchens and pantries in search of sweets. They can also bring infestations of root aphids to your ornamentals and are frequent invaders of greenhouses. The workers start appearing in July and are around all summer.

Control: Ant control begins with careful sanitation and good housekeeping. Make sure that foods are kept in sealed tins or glass containers. Pick up crumbs and bits of food that will attract ants. Of course if there are children and adolescents in the house, keeping food picked up will be impossible, so assume you will have to resort to chemical control.

Minor infestations can be partially controlled by use of commercially available ant baits or traps. These baits represent a form of slow attrition, since all they do is knock off some of the foraging workers (of which the nest provides an unending supply), so the best way is to go to the source, the nest.

Indoors: sprays, dusts or granules containing diazinon, chlorpyrifos, malathion or propoxur are effective. For quick knockdown of ants, house sprays containing pyrethrins will do the job. Indoors: oil- or water-based sprays are preferable because they dry to an almost invisible coating. Outdoors: water-based sprays, mixtures of wettable powder or emulsifiable concentrate (agitate the mixture constantly to prevent settling out) can be used. Carbaryl is registered for outdoor use. Indoor infestations of apartment buildings or row houses require cooperative efforts.

To prevent ants from entering, spray all around the foundation of the house, being sure to follow directions on the label. In the house, spray cracks, moldings, baseboards, window- and doorframes, cracks, areas around electrical outlets and plumbing. Sometimes you can use a small paintbrush to apply the pesticide. Repeat every two months. Outdoors, pour the pesticide into nest openings and over the surface of areas around the openings.

To control ants in the home, on lawns or in gardens, try to find their nests. Note the surfaces over which ants crawl as they move into or through the house. This will help you to determine whether the nests are inside the house or outside. Next, apply an insecticide (pesticide) to the nests, if you can reach them, and to the surfaces over which the ants crawl.

By applying a long-lasting insecticide to the surfaces, you can prevent new infestations after you have rid your home of the present one.

Although most species have certain preferences for nesting places, knowing the kind of ant you are trying to control does not always help in locating the nests. Ants are highly adaptable in their nesting habits.

For example, pavement ants usually nest along sidewalks and driveways and around or in foundation walls, but they also may nest in the basement, under a porch or in the lawn.

Indoor-nesting species sometimes nest outdoors during summer in the North and the year around in the South.

Outdoor-nesting species sometimes make their nests inside a building. You can usually find their nests by watching the movement of the ants. Note where they are coming from or where they are going after feeding.

Finding an ant nest inside the house is often difficult. It may be impossible without removing a wall or floor. Ants that nest in the house usually must be controlled without locating their nests.

You may trace ants to an electrical outlet or to the area back of a baseboard or door frame, but their nest may be some distance away—perhaps in the framing of the house.

Indoor ants may nest between the floor and subfloor, in the walls, behind baseboards, beneath cracked basement floors, in decayed or rotting house timbers, or even in a pile of papers.

To find the nests outdoors, you may have to do no more than look for anthills on the ground. When some ants build nests in the ground, they thrust out bits of earth, which form anthills around the openings.

Some nests are a little difficult to find because they are under walks or driveways (brick, flagstone or concrete), under boards or stones, or next to foundation walls. Nests also may be found in decaying logs or tree trunks.

BEETLE PESTS OF STORED FOOD

Some of the beetles that can become irritating annoyances to the homeowner are occasionally introduced into the kitchen or pantry by means of infested food packages. These pests do little real harm, but do munch holes in packages, contaminate foodstuffs with their droppings and startle you when they scurry about your shelves. They infest all manner of grains, beans, dried fruits, nuts,

Saw-Toothed Grain Beetle

Larva   Adult

Granary Weevil   **Rice Weevil**

pasta and candy. In this section we will cover only the more common of these pests:

The granary weevil, *Sitophilus granarius*, and the rice weevil, *Sitophilus oryza*, are common long-snouted beetles that feed on a variety of grains. They are really quite similar in appearance, with the rice weevil limited to the warmer, southern parts of the country and the granary weevil limited mainly to more northern climes. Adult granary beetles are shiny red-brown, up to a quarter of an inch long (usually less) and have long snouts or beaks. The rice weevils are much smaller (one-tenth of an inch long) and have four light red spots on their backs. The female weevil gnaws into individual kernels of grain and deposits an egg; then seals the hole. This process is repeated again and again. The eggs develop into small yellow-white grubs that feed on the grain for a week or two before pupating.

Another common pest of grains, cereals and other foodstuffs is the cadelle ("bread beetle"), *Tenebrioides mauritanicus*, a shiny black, nocturnal, secretive, long-lived beetle, whose ravenous gnawing larvae do enormous damage. The larvae may live for more than a year, feeding the whole time on anything they find, including cigars. They gnaw holes in packages and can even burrow into wood, books and carpets. Similarly shaped though smaller (one-tenth of an inch in length is the very common pest, the brown drugstore beetle *(Stegobium paniceum)* which feeds on practically anything, including books. This pest is common in homes and feeds on any edible material, including spices such as

Drugstore Beetle

Larva   Adult

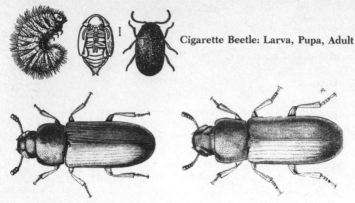

Cigarette Beetle: Larva, Pupa, Adult

Red Flour Beetle   Confused Flour Beetle

red pepper and paprika. Its boring is not limited to foodstuffs but includes wood and books. These brown, cylindrical, tenth-of-an-inch-long adult beetles retract their legs and antennae when not roaming about, so they are not particularly obvious. Like other beetles, they lay their eggs in foodstuffs; the eggs hatch into larvae, which spend the next four months munching away.

The cigarette beetle, *Lasioderma serricorne*, is a reddish-brown, ⅛-inch-long, oval-shaped pest that, like the drugstore beetle, thrives in warm buildings. As the name implies, it bores into tobacco and can devastate a box of very expensive cigars in short order. These beetles are not all that choosy and will feed on a wide variety of household foods and spices, as well as wool, paper and leather.

Flour beetles of the genus *Trilobium* are tiny beetles whose history goes back to ancient times. They probably originated in the tropics and do well in warm climates. These pests are reddish-brown and only about a tenth of an inch long; hence they can get into containers that aren't tightly sealed. They get into all manner of grains—flour, bran, etc.—and cause these substances to turn gray, moldy and smelly.

Mealworms of the genus *Tenebrio* are nocturnal, half-an-inch-long, dark brown to black beetles that are found throughout the United States. They lay their eggs in grains; the eggs develop into large, worm-like grubs (dark brown or yellow, depending on the species) that are up to one and one-quarter inches in length. Many people actually raise mealworms to feed their tropical fish. These larvae not only infest foods but also actively wander about and burrow into all manner of things.

There are numerous other beetles that exhibit kinds of behavior similar to those described above. A detailed review of each of these would add little to this book, since the tactics and strategies for dealing with all of them are the same.

Control

If you discover a beetle infestation in your kitchen or pantry, immediately get rid of the infested material by incineration or other means. Clean out your cupboards by removing all food items and, after washing down the shelves, spray them, including cracks and crevices, with insecticides that contain malathion or propoxur. Make sure you do not spray food preparation surfaces. After the spray has dried, reline the shelves with paper and put all foods, such as cereals, grains, pasta, chocolate, dried fruit, nuts and spices into tightly sealed glass or metal storage containers. To prevent reintroduction, don't buy damaged packages whose seals have been broken. Some foods that attract these pests can be stored in paper packages so long as they are stored in the refrigerator—most of these beetles have tropical origins and don't survive refrigerator temperatures.

MOTHS THAT INFEST GRAINS

There are several moths, Lepidoptera, that are pests of kitchen products such as flour, dried fruits, crackers, beans, chocolate and nuts. The most common of these are: the Indian meal moth, *Plodia interpunctella*, a ⅝-inch-long moth with bronze wings tipped with a broad gray band, and the Mediterranean flour moth, *Anagasta kuehniella*, a night-flying, inch-long pest with gray-white banded wings. These moths lay their eggs on any available food, and the eggs develop into half-inch-long, whitish-to-pink larvae. The larvae have silk glands and spin tubes of silk continuously. This behavior causes flour to stick together in clumps.

Control: Get rid of contaminated dried fruit or flour, and then thoroughly clean out the cupboard. If the infestation is heavy, spray shelves with pyrethrin- or methoxychlor-containing insec-

Indian Meal Moth Larva

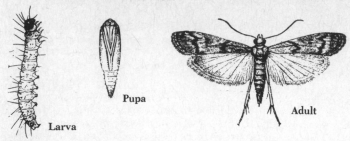

Mediterranean Flour Moth

Larva

Pupa

Adult

ticides. Let the spray dry, line the shelves with paper and make sure all possible edible foods are stored in tightly sealed plastic, glass or metal containers. Be careful to inspect packages of food when you buy them and reject any packages that are damaged or whose seals are broken.

HOUSEFLIES

One of the most annoying and common of the insect pests of the home are the houseflies: *Musca domestica* (the housefly), *Phaenicia sericata* (the greenbottle fly), common bluebottle flies of the genus *Calliphora* and *Phormia regina* (the black blowfly). These flies are somewhat similar in appearance, having the same basic form but with color variations. The housefly is up to a quarter-inch long and has black stripes on its thorax and abdomen. The black blowfly is about a third of an inch long and is uniformly dark blue, the bluebottle flies are about 2/5 inch long and have metallic blue abdomens, while the greenbottle fly is about a quarter-inch long and has a metallic green sheen.

The three houseflies and a number of related species require fresh, fermenting animal excrement or decaying animal tissue or garbage to breed. Females deposit their eggs in such materials, and these hatch into maggots in less than one day. These disgusting, creamy-white, footless, greasy grubs reach their maximal size, one-third of an inch, in a week or more, depending on the

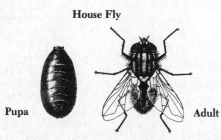

House Fly

Pupa

Adult

temperature, and then move to a cool, dark place to pupate and eventually metamorphose into adult flies. The adults, depending on available food, live several weeks. They are strong fliers and may fly several miles from their hatching site at heights of about 80 feet. Most are killed by the winter cold, but enough pupae make it through the winter to provide a new generation of nuisance pests when the weather warms up.

Flies at times are unbearable pests as they buzz noisily about, but the real problem with flies is that they are dirty and often feed on feces, in the process accumulating all manner of disease organisms on their hairs and bristles. Among the diseases known to have been transmitted to humans by flies are cholera, dysentery, typhoid fever, tuberculosis, anthrax and salmonellosis, a common cause of food poisoning.

Control: The best way to keep flies out of the home is to prevent their entry by well maintained screening, 18-mesh or smaller. Even this size screening will not keep out some of the smaller biting flies, midges and no-see-ums (using a mesh small enough to exclude these little pests would reduce air circulation unduly). To counteract these small flies without compromising the air circulation, screens can be sprayed or painted, at six-week intervals, with 6% deodorized malathion or 6% propoxur. Another strategy has been to spray the screens with repellents such as deet (see page 69). However, their residual effect is relatively short.

In the home, control of flies can be gained by a combination of space sprays with rapid knockdown effect and residual surface sprays. Many commercially available pressurized contact poisonous mists are available. For these sprays to work, droplets of the poison must hit the target insect, but the results are rewarding, except for big horseflies, and provide fast (but only temporary) control. Pressurized cans are expensive and cannot be incinerated; moreover the propellant may have negative effects on the environment. A cheaper solution is to use hand sprayers, although the droplets formed by these tend to be bigger than those from pressurized cans. For larger areas, such as sheds, garages and barns, pressurized foggers are effective. Most fly sprays contain pyrethrum or pyrethroids, which provide rapid killing and are of relatively low human toxicity. Surface or residual sprays provide longer-term protection but relatively slow knockdown. They are applied as sprays to moisten windowsills, light fixtures, walls, ceilings and screening; when they dry, they leave a coating of poison which will get on the flies. The most effective residuals, such as DDT, are now banned because of their potential for detrimental effects on other animals, including humans. Unfortunately, many flies, particularly the housefly, have developed resistance to many of the current insecticides.

Some commercially available hanging fly killers that emit toxic vapors are also effective but should not be used in kitchens or dining areas. Sticky flypapers also work, but are far from being *objets d'art*. Many people have also invested in electrical devices that attract insects to an electrically-charged grid that electrocutes them on contact. These devices are expensive, not particularly effective, unappealing and—worst of all—emit a loud "zap" even when dispatching a tiny midge.

Finally there is the old-fashioned flyswatter which, if you have patience and a little fine motor control, provides you the satisfaction of the hunt, revenge on your tormentor and an immediately dead fly. There is no greater satisfaction than clobbering two or three flies with one well-aimed stroke. Some people prefer the backhand, others the forehand and, for the more vigorous, a full overhead will get the job done.

FRUIT FLIES

Fruit flies of the genus *Drosophila* are small flies that not only are occasional annoying pests around food and fruit but have also been valuable participants, albeit unwilling ones, in much of our early research on genetics. Indeed, the favorite laboratory fly, *Drosophila melanogaster*, is the one that occurs most frequently in the home. These ⅛-inch-long, bristly flies have tan heads, black abdomens and bright red eyes. They are attracted to decaying materials, most usually fruits or vegetables, and fermenting fluids such as cider in poorly sealed containers; hence one of their common names is "vinegar fly." They lay their eggs on decay spots, and these soon (in 30 hours) hatch into small maggots that feed for about a week, then become pupae. These will emerge as adult flies, quickly becoming sexually mature and starting a new generation. Thus their numbers can increase rapidly and they can become an unattractive nuisance.

Control: No control measure other than good sanitation is called for in getting rid of fruit flies. Seal containers, keep fruit in the refrigerator, don't put food-contaminated mops, brooms or cleaning rags away without washing them. Be sure to clean out cracks and crevices that might contain fermenting food products. Also, check around drainpipes under sinks and rinse out your trash containers frequently. Use plastic, sealable trash and garbage bags.

Cockroaches

About 250–300 million years ago, during the dank, wet Carboniferous period, cockroaches reached a point in their evolution that was so successful that they persist in pretty much the same form today. They are truly extraordinary creatures that are so well adapted and prolific that, despite all human technological advances, we can not get these pests permanently out of our lives and homes. In urban areas, more money is expended annually on attempts to control them than is spent fighting any other pest. Their love affair with human habitations goes back to the development of heated homes; while they are not as destructive as their distant relatives, the termites, they are universally viewed as loathsome pests. The damage they do to paper and fabrics is not great nor are they a major health threat, although they may contaminate food with their droppings, spreading disease-causing organisms. In tropical areas where roaches abound, there have been reports of them munching on fingernails of sleepers and even nipping off the eyelashes of sleeping infants. In one recent horror tale, a prisoner incarcerated in a roach-ridden jail brought legal action against the state because a roach burrowed into his middle ear and caused him great mental anguish until it could be surgically removed.

The main problem with these uninvited house guests is their very presence. They leave their messy droppings all over the place. If there are lots of them there is a distinct fetid stench that announces their presence to one and all. They are everywhere—kitchens, bathrooms, under sinks, refrigerators, in household appliances, in every crevice and crack, in cupboards, drawers, closets, cabinets and pantry shelves. Although several of the common household species can fly, they usually scurry about rapidly on their long legs. Most are stealthy, nocturnal pests that scuttle for cover when a light is turned on. Thus, the Romans gave them the name *lucifuga*—"those who flee from light." The name *cockroach* has its origin in the Spanish word *(la) cucaracha,* and there is a rather energetic piece of music that has a beat which captures the rate of its movement. The human response to their presence may be in part due to their sneaky behavior and in part to their appearance, which is not as loathsome as that of some other insect pests, but nonetheless does evoke shudders from some people. They are not particularly large: although the American roach may reach an inch and a half in length, most are smaller. They are flat, shiny and oval-shaped, have long, thin antennae and vary in color from golden tan to reddish-brown to black. They are very prolific; the introduction of a single egg case into a home could produce a real infestation in a matter of several months. Maybe most of us dislike roaches so much because we can't beat them. The best that can be done is to control them—and this requires a special effort.

While there are some 4000 species of cockroaches in the world, only a dozen or so frequent human habitations. The four most common household roaches are described in the following pages.

The German Cockroach

The German cockroach, *Blattella germanica,* is found all over the world. These creatures are pale brown, winged, boat-shaped, half-inch-long pests whose reproductive potential is enormous. The female, usually darker than the male, has a fatter, more rounded rear end than the male, lives up to 200 or 300 days and will produce in that time four or five egg cases, each containing about 50 eggs. These egg capsules, which may become as large as the female's abdomen, remain attached to her back end until hatching—in about two weeks—when the egg case splits and the larvae make their exit. Reproduction is best at warm temperatures (70° to 80°F) and when there is a relative humidity greater than 40%. The larvae go through a series of seven gradual developmental stages (nymphs) during which they actively run about and feed. Under ideal conditions of temperature, humidity and food availability, these roaches reach adulthood in about 60 days, but under less than ideal conditions the process may take 90 days or more.

Both adults and nymphs hide by day and prowl by night, usually not far from their hiding places. They prefer dark, moist areas near their food sources, which are highly variable. They eat almost any kind of organic materials, particularly starchy materials. They are particularly fond of beer (perhaps part of their Germanic tradition), and bread soaked in beer is an ideal roach bait, although

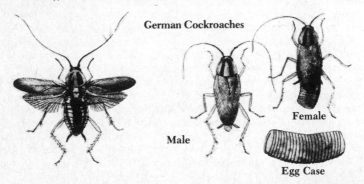

German Cockroaches

Male

Female

Egg Case

they will be attracted to almost any fermenting food. Because of their daytime secretiveness, one is sometimes surprised by the enormous numbers that are flushed out when spraying into their hiding places. They tend to be gregarious, forming large colonies and sometimes, when their populations get too large, will migrate in search of more fertile hunting grounds. In severe infestations, tens of thousands of roaches can be present in a dwelling. Indeed, the German cockroach is the most common of the house roaches.

The American Cockroach

This large (up to 1½ or 2 inches in length), reddish-brown, moisture-loving roach is commonly found in homes, usually in basements, around pipes and in sewerage systems. Its affinity for water accounts for one of its common names, waterbug, but its real name is *Periplaneta americana*. In Florida, these large, winged flying pests are called "palmetto bugs"; in England, dockworkers, finding hordes of them on ships arriving from India, dubbed them "Bombay canaries." They can be identified not only by their large size but also by the light yellowish markings on their pronotum (front segment). The female carries her egg case only briefly, and then glues it to some well-protected surface, where it sits for 40 to 50 days before splitting open and releasing about a dozen white nymphs. These go through a protracted gradual development period, molting 13 times over the span of about a year or more before they reach adulthood. Females produce egg cases throughout the year, but most rapid production is in the summer months. The females have a life span of a year or two and may produce 60 or more fertile egg cases. The American roach gets along well with German roaches, and some homes may have simultaneous infestations of both types. The American roach also likes beer, so rinse out your empty beer cans before storing them for recycling.

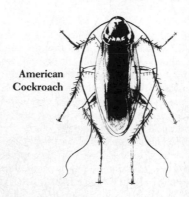

American
Cockroach

Brown-Banded
Cockroach

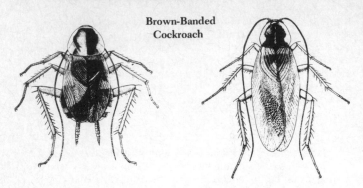

The Brown-Banded Cockroach

A frequent pest in many southern cities is the highly gregarious, brown-banded roach, *Supella longipalpa*. It is thought to have entered the United States in Miami around the turn of the century. It now can be found scurrying around cupboards and pantries and even upstairs rooms in areas where the temperature reaches 80°F or more. This roach is about the same size as the German roach, adults being half an inch long, and having two tan transverse lines at the base of its wings. The female produces compartmentalized, yellowish egg cases containing a dozen or so developing nymphs. The egg case is carried for about a day and then glued to some protected, dark surface. Seventy days later, the newly hatched nymphs emerge. Development is gradual, and about 70 days after emerging, they reach adulthood. This pest has hitchhiked its way northward, and is now widely distributed, but still is commonest in the warmer parts of the country, particularly the South.

The Oriental Roach

The highly despised Oriental cockroach, "waterbug" or "black beetle," *Blatta orientalis*, is a dirty, sluggish, bold roach that travels through sewer pipes and favors damp basements. Often they spread through apartment houses, making their way up pipes or incinerator shafts to the upper floors. The inch-long, very dark brown adults come in two forms: winged males and almost wingless females. The female has a rather wide abdomen that almost drags on the ground as she walks along. There are usually more females than males in a population. Reproducing females carry an egg case projecting behind them for a few days before depositing it in some warm, dark protected place. The compartmentalized egg case sits for about 60 days before hatching out most of

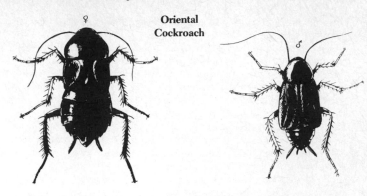

the fertilized eggs, from which small, pale-ish, wingless nymphs emerge. Over the next 12 months (and seven moltings), the nymphs become sexually mature adults that live for up to six months. These roaches are quite gregarious, and sometimes are flushed out in great numbers when sprayed with a pesticide.

Roach Control:

Good Sanitation: The first steps in control are: (1) good house-keeping; (2) denial of entry and (3) dryness. Roaches are omnivorous, so complete denial of food is difficult, but leaving opened containers or bits of food about not only is an open invitation for these pests but also insures that they reproduce under optimal conditions. Keep trash and garbage in covered containers, sweep and vacuum frequently, don't leave food in your pets' dishes, empty litter boxes frequently and always rinse out beer containers—magnets for roaches. Roaches often gain entry through cracks or through openings where plumbing or electrical connections come through walls. Deny access by calking all such openings.

Chemical Control: We humans like to think all problems can be resolved, but the probability of eliminating roaches is about the same as that of achieving a lasting peace in the Middle East. Just when you think you've got the problem licked, it suddenly crops up again. However, don't despair; you can achieve good temporary results and can limit the degree of cockroach infestation with a variety of chemical weapons. In other words, you can *manage* roaches with pesticides, with this proviso: these pests, because of their genetic plasticity and huge reproductive potential, may in time develop resistance to a given poison. You will then have to change poisons.

There are a number of sprayable, residual (long-lasting) contact pesticides that provide weeks or even months of control. Among these are acephate, bendiocarb (a carbamate), carbaryl, chlorpyrifos (which unfortunately has a foul, sulfurous odor), malathion (also smelly), diazinon and fast-acting propoxur (also a carbamate). In addition to residual sprays, contact poison dusts blown into hard-to-reach cracks, crevices, nooks and crannies may afford long-term protection. Among these dusts are the old-time roach remedies, borax powder and boric acid powder, which will not give the quick results seen with propoxur; nevertheless, these are valuable in a long-term pest management program. Like all toxic agents, even boric acid is potentially toxic to humans, particularly children, so don't leave the container about and use the dusts in places inaccessible to kids. Other slow but effective inorganic dusts are silica aerogel and sodium fluoride. These inorganic dusts may be used in combination with organic pesticide dusts such as long-lasting diazinon and bendiocarb and short-acting, quick-knockdown pyrethrins.

In addition to spraying and dusting another level of control can be achieved with the use of poisoned baits placed wherever roaches are seen. These baits contain a variety of different toxins such as chlordecone, chlorpyrifos and propoxur. They are commercially available under a variety of trade names such as Pest-B-Gon. Boric acid baits are also available. There are also sprayable and paintable lacquers and poisonous tapes that are commercially available. For those of stout heart, a recent report from Florida described a very effective roach control method. A couple there brought into their apartment a dozen large banana spiders, and within a few weeks there were no more roaches.

Having thoroughly dealt with the negative aspects of roaches, now for something positive: they eat bedbugs.

Roaches are the ultimate survivors. They can tolerate one thousand times more ionizing radiation than man. So if it is any consolation to you, be advised that whatever creatures take over the world after the nuclear holocaust, they too are going to have to deal with cockroaches.

SILVERFISH (BRISTLETAILS)

Over 300 million years ago, insects of the order Thysanura first made their appearance on this planet. These thysanuran primitive ancestors gave rise to a sort of living fossil, the present-day fabric, starch- and paper-eating, nocturnal pests called silverfish.

They are frequently seen scurrying for cover in kitchens when the light is turned on by late-night snackers. They are often found also in smooth-walled bathtubs and sinks into which they have tumbled during their quest for food. Contrary to popular belief, they did not come up the drainpipe, but they do follow their evolutionary past and prefer areas with high moisture content. Hence they are most frequently seen in bathrooms, basements and kitchens for, despite their ancient past, they have adapted very well to the opportunities presented to them by human dwellings.

Silverfish are shaped like elongate teardrops, being wide in the front and narrow toward their rear ends, from which project two short and three long, segmental, anal appendages called cerci. Because of these cerci, some call them "bristletails." At their front ends, each has a long pair of slender, rearward-curving, segmented antennae. They have three short pairs of legs, no wings, and are covered with scales. Silverfish have long life spans—up to five years—and unlike most insects, produce only a few offspring, about 20 in a lifetime. The eggs are deposited in crevices and hatch into miniature look-alikes of their parents.

While these pests don't eat much and prefer starches, they do have the capacity to digest cellulose; consequently, they can damage books, papers, stamp collections, wallpaper and the like. They

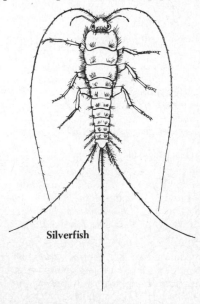

Silverfish

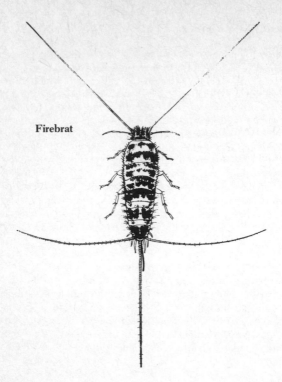

Firebrat

love meat and are occasional cannibals. Their sweet tooth attracts them to kitchens, and they are sometimes found in flour and cereal containers. These omnivorous pests also munch on starches and fabrics made from plant fibers, but they leave silk and wool alone. They have a considerable range in their quest for food and often leave announcements of their presence in the forms of yellow feeding stains, dark, tiny droppings and shed scales. Frayed edges on papers and textiles give warning that silverfish have been feeding.

There are several species of silverfish that have become household pests. The half-inch-long, gray, dark-marked firebrat (*Thermobia domestica*), often found near furnaces and heaters, the ⅜-inch-long four-lined silverfish (*Ctenolepisima lineata*), the half-inch-long true silverfish (*Lepisima saccharina*) and the ¾-inch-long gray silverfish (*Ctenolepisima longicaudata*).

Control: Start your silverfish control program by eliminating moisture (leaking pipes, etc.) and depriving them of an ideal habitat. Seal up cracks and crevices in areas where they may nest and hide by day, clean up food scraps and pet food trays and keep starchy foods in sealed containers.

Quick knockdown sprays (e.g., pyrethins) will flush out and kill silverfish, but often in their rapid flight from the spray they may escape a killing dose; thus, such chemicals should be used in combination with long-lasting, sprayable contact poisons, such as 0.5% chlorpyrifos, propoxur or diazinon, 0.25% bendiocarb or 1% malathion. Dusts similar to those described for roaches (see page 131) provide a measure of control, as will baits placed at strategic locations for example at corners of book shelves, underneath heaters and near cracks and crevices wherever silverfish have been seen.

BOOKLICE, OR PSOCIDS

Grain and grain products are sometimes found to be swarming with minute insects that are scarcely larger than a pinhead. Flour and grain samples appear to be particularly attractive to them. They are known as psocids, or booklice, owing to the resemblance to lice and to their occasional presence in books. They belong to several closely related species of the genus *Liposcelis*. They are tiny, pale-grayish or yellowish-white, wingless, soft-bodied louse-like insects with fairly large heads, poorly developed eyes and long, slender antennae. They are about one twenty-fifth of an inch in length. They feed on a great variety of organic matter, both of plant and animal origin, but are troublesome through their presence rather than for the actual damage they cause.

In some cases females are believed to reproduce without mating; in other cases males have been found. As many as one hundred eggs may be laid by a single female, and in summer the developmental period from egg to mature insect is about three weeks. The newly hatched young resemble the mature insects in form and general appearance but are smaller and of a lighter color. These insects are widely distributed in North America and Europe.

PART V

Pests of Fabrics, Furs, Leathers, Carpets

CLOTHES MOTHS

There are a number of small moths whose larvae attack and damage clothing, rugs, draperies and other household goods containing animal fibers, wool, furs or even feathers. These pests have been with humans for thousands of years and are even mentioned several times in the Bible. Fortunately, in the past two or three decades, they have become much less of a pest than in the past, partly due to the increased use of synthetic fabrics, and partly because of better and more powerful vacuum cleaners and dry, forced hot air heating.

Clothes moths have the capability of digesting the almost indigestible protein keratin, which is the main constituent of wool, hair and feathers. Given this capability, they do considerable damage, but just as man cannot live on bread alone, moths can't live on wool alone, since wool lacks certain nutrients essential for their survival. This being the case, fabrics stained with food or sweat are particularly attractive to moths since they contain nutrients that are necessary for them to thrive. One way to prevent moth damage is to always have your winter woolens thoroughly cleaned before putting them away for the summer.

There are two moths that are the usual culprits in attacking fabrics: the casemaking clothes moth and the webbing clothes moth, the latter being a much more serious pest.

The Webbing Clothes Moth

The adult webbing clothes moth (*Tineola bisselliella*) may be found in heated buildings all through the year in almost every part of the temperate United States and Canada. The adults are small, with a wingspan of about half an inch, and have narrow wings covered with dull, golden-brown scales. Unlike most moths, they are not attracted to light, and when disturbed, rapidly hide in some dark, secluded place. They do best in a warm, moist environment, and hence do considerably more damage in the southern than in the northern United States. Although they don't attack silk or synthetic fibers, they do go after clothes made of a mixture of wool and synthetics. After mating, females deposit up to one hundred tiny white eggs on clothing or rugs. Then they die. The eggs are glued to the threads of the target object and hatch in about a week. The newly hatched, very active larva is about 1/25 of an inch long, white and translucent. It begins feeding at once while spinning a fine silken tube about itself. Later, as the larva grows, the tube becomes covered with debris. Sometimes the larva leaves its tube to feed, but most of the time it confines its feeding to fibers around the tube. Larvae may live up to a year or more and attain lengths of almost half an inch before spinning a silken cocoon and pupating (usually in the early summer). About a week later, adult moths emerge and start a new cycle of mating and egg-laying. However, in heated homes there can be as many as four generations in a year.

The Case-Bearing Clothes Moth

The adult case-bearing clothes moth, *Tinea pelionella*, looks very much like its cousin, the webbing clothes moth, except that its scales are browner and it may bear dark spots on its wings. It is not very common but will do better in northern climes and lives outdoors, often associated with birds' nests. The brown-headed larvae of this pest are easily recognized because they spin ⅓-inch-long tubular cases that they drag along as they feed. When disturbed, the larva pulls back to the security of its tube. It attacks down-filled upholstery and all manner of woolens and furs, and then pupates, only to emerge as a sexually mature moth during the summer.

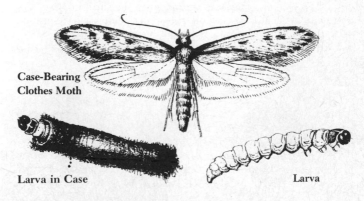

Case-Bearing Clothes Moth

Larva in Case

Larva

Control: As is always the case, an ounce of prevention is worth a pound of cure, so control of clothes moths begins with good cleaning. Woolens put away in storage should always be dry-cleaned first. Washables should be washed and ironed to get rid of any of these pests. The clean clothes should then be put away in tightly sealed plastic bags. Stored clothes can also be protected by mothballs or mothflakes containing napthalene or paradichlorobenzene. These volatile chemicals readily vaporize and act as gases that kill the destructive larvae, adults and eggs, but be warned: you must use enough (follow instructions on the label), and they only work in tight containers that will keep in the fumes. New cedar wood also gives off a potentially toxic vapor, but cedar closets and cedar chests lose their zap after a few years and afford little or no protection.

A number of pesticides are of value in controlling a moth infestation. These include pyrethrin (0.5%), resmethrin (0.25%) and other synthetic pyrethrin-like chemicals. They may be sprayed on clothing (make sure to read the label) until the fabric is slightly moistened. Also spray storage areas, closets, chests, shelves, floors, baseboards, under carpets and along the margins of carpets. Protection afforded by these chemicals is of limited duration—only a few months—so periodically check your storage areas.

BEETLES THAT ATTACK FABRICS AND LEATHER

The larvae of some beetles of the family Dermestidae feed on woolens, furs or other materials containing animal protein. These destructive pests find their way into your home in a number of ways. Some, like the carpet beetle, are pollen feeders and find entrance on cut flowers; others are commonly found in mammals' or birds' nests near or in the home, while some, ordinarily pests of packaged foods, may find entrance via the shopping bag. For the purposes of this discussion, the carpet beetles will be used as an example, since they are the most important pest species.

Carpet beetles are small (usually less than a fifth of an inch

Common Carpet Beetle

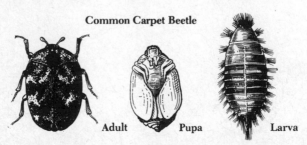

Adult Pupa Larva

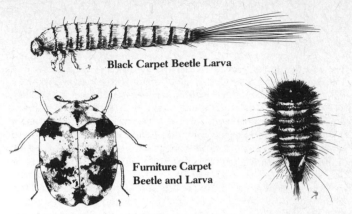

Black Carpet Beetle Larva

Furniture Carpet
Beetle and Larva

long), rather attractive beetles, somewhat oval in shape and possessing short, knobbed antennae. The common carpet beetle, *Anthrenus scrophulariae*, while widespread, is found mostly in the northern and midwestern states. As its name implies, it is a special pest of fur rugs and wool carpets. The adults are black and have white specks and an orange band down the middle of the back. Eggs are anchored to fabrics or furs and hatch into brownish, very active larvae covered with hairs. The larvae feed for about two months and then pupate for a couple of weeks before giving rise to a new generation of adults. The black carpet beetle, *Attagenus megatoma*, is a widespread, highly destructive pest that is a major threat to homeowners in the eastern part of the country. The adults are a tenth to a fifth of an inch in length and very dark brown in color. They lay fragile eggs that develop into slender, hairy, golden brown larvae that burrow into the nap of rugs and wreak havoc for a year or more (depending on the temperature and humidity) before pupating. In the south and southwest, the furniture carpet beetle, *Anthrenus flavipes*, attacks hair-filled upholstery, rugs, furs or any other object containing animal protein.

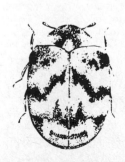

Varied Carpet
Beetle and Larva

The black adults are about a fifth of an inch long, oval-shaped and mottled with yellow or white spots. Eggs hatch into larvae in about two or three weeks. The larva has a long tuft of hairs at its rear and feeds voraciously for up to a year before pupating.

In the western states, the main dermestid household pest is the varied carpet beetle, *Anthrenus verbasci;* this one is blackish and mottled with white or yellow spots. Its striped, hairy larva may ravage fabrics and furs for a year or more before pupating.

Some of the beetles described as kitchen pests may also attack fabrics, leather and furs as well as foodstuffs (see page 138), but all are treated in more or less the same way.

Control: A number of commercially available insecticide sprays that are used in controlling clothes moths (see page 138) are also effective against carpet beetles. These preparations may contain bendiocarb, chlorpyrifos or dichlorvos (DDVP). Read instructions carefully and spray around rug edges, under carpets, on floors, in closets, on shelves, on baseboards, etc. Bendiocarb can be used to treat the whole rug and has some residual action, but, since all these pesticides will lose potency in time, they should be applied at intervals prescribed on the label. In severe infestations, it may be necessary to call a pest control organization with fumigation facilities. Valuable furs and leathers, when not in use, might best be protected by putting them into a commercial cold storage facility.

CRICKETS

We have all listened to the noisy, shrill chirps of crickets on a warm summer night and, if you are an astute observer of the world around you, you are already aware that the chirping rate goes up as the temperature increases. Indeed, crickets can be used as Fahrenheit thermometers by adding 40 to the number of chirps in 15 seconds. The song of the cricket is an announcement of the heat of passion and is played only by males. The chirping sound is produced by a rasp-like stridulating organ set between its two wings. The sound is not unlike that made by twanging the teeth of a comb; females are either attracted to or turned on by the most proficient chirper. In the Orient, ornate cricket cages were part of the household decor, and in the pre-stereo days, the crickets filled the silent rooms with sound. I must say their song is preferable to some of the super-amplified excuses for music that my children blast out of their stereos, but whereas stereos can be turned off, a cricket in your closet can keep you up all night. Actually, field and house crickets do relatively little damage when

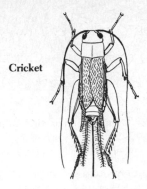

Cricket

they invade your home—possibly a few munches out of some favorite piece of clothing. The large-headed, bizarre-looking Jerusalem cricket might possibly take a bite out of you, and the burrowing mole cricket may make holes in your lawn, but otherwise they are just noisy, uninvited guests that occasionally invade in large numbers.

Crickets belong to the insect order Orthoptera, and thus are close relatives of grasshoppers. They have enlarged, powerful, long jumping legs and seldom use their wings, preferring a jumping mode of locomotion. The common house cricket (*Acheta domestica*) is a little over half an inch long and is colored light yellow-brown with three prominent darker bands on its head. It has long, thin antennae, six legs and a taste for silk and wool. Crickets are nocturnal hunters and are marvelous at playing hide-and-seek. I recall spending an hour trying to find a lovelorn male in my bedroom closet. Never did find the noisy little musician. President Reagan and his wife Nancy were kept up all night by a single cricket that was able to avoid the best efforts of White House Security.

Control: I can't get too excited about cricket control unless there is a massive invasion. You can deny them water by repairing leaky pipes (a good idea even if you don't have crickets), and of course you should not invite them in by leaving food debris about. Residual contact sprays should do the job (see section on roach control, page 131).

CLUSTER FLIES

In the fall, with the approach of cold weather, cluster flies (*Pollenia rudis*) begin to seek a warm place to spend the winter. Unluckily,

Cluster Fly

their winter vacation ground is your home, into which these me-
dium-size, dull-gray-black intruders sneak one by one. They get
their name, cluster flies, because of their behavior (quite sociable).
When they find a nice dark nook or cranny, they like to get
chummy there with a crowd of friends. Many a homeowner is
shocked to take down a hat from his closet and find a fair-sized
gray bunch of torpid, semiconscious flies. Worse yet: on warm
days they are aroused and swarm to sunny windows.

Cluster flies do no real harm, but they are uninvited guests
that intrude in our space, and for some people are a real nuisance.

Control: The best thing to do about cluster flies is to prevent their
entrance with good screening, making sure to winterize your air
conditioning vents to the outside. Once they get into the house,
commercially available fly sprays and vacuuming will take care of
any immediate cluster fly problems.

PART VI

Houseplant Pests

PREVENTING INFESTATION

For many people, ornamental plants have become a major part of their homes' decor. These attractive and often expensive plants can be ravaged by a variety of insect pests, causing you the loss of the plant and in many cases anxiety that the pest will spread to your other plants, as well it might. Thus, the first step in maintaining healthy household plants is to protect against possible infestation. Protection begins with the plant purchase. Examine all plants for any evidence of insect pests. Mealy bugs, whiteflies, aphids and scale insects are easily detectible by the naked eye, and this book will provide you with guidelines of what to be on the lookout for.

Even apparently uninfested new plants can harbor eggs or a few hidden pests whose reproductive capabilities can produce an infestation in short order. Therefore, the second protective step is to quarantine or isolate all new plants for two to four weeks and keep them under observation. It would be a shame to have the insect plague spread to all your plants. Sometimes pests are brought in with cut flowers, so it is a good idea to inspect and carefully wash off cut flowers brought into your home.

Despite our best efforts to prevent infestations, sometimes these insidious pests do get into our homes and greenhouses. Like all infestations, they start out as a minor problem that can be solved easily by a variety of simple strategies involving physical rather than chemical control. Make it a habit to examine your houseplants when you water them, because the sooner you catch an infestation the greater the ease of control.

PHYSICAL CONTROL OF MINOR INFESTATIONS

If you see evidence of an infestation, isolate the plant and treat it individually, preferably outdoors. Wash the plant with soapy water, two teaspoons of mild detergent per gallon of water. To apply the soapy water, use a soft cloth, sponge or soft brush. Then spray the plant down with a hose or spray nozzle in the kitchen sink. Another tactic is to put a bucket of soapy water in the bathtub, turn the pot upside down and immerse the entire plant in soapy water. This strategy will often remove most of the aphids, mealy bugs and scale insects from broadleaf plants.

If there are only a few bugs, cutworms or slugs, hunt them at night by turning on the lights. They can be picked off with tweezers and flushed down the toilet. Another effective method in minor infestations is to take cotton swabs soaked in ordinary rubbing alcohol and use them to pick off individual insects. These methods repeated from time to time should control the pests. If the infestation is severe, the best decision is to get rid of the diseased plant. It is sometimes penny-wise and dollar-foolish to try to save a badly infested plant that can spread the pest to the rest of the plants in the house. When getting rid of infested plants, put the whole plant in a plastic bag, seal it tightly and put it in the trash. Never simply dump it outside, where it can provide a reservoir of pests that can be re-introduced. Should these physical methods fail, then you can begin to use biological and chemical control methods.

CHEMICAL AND BIOLOGICAL CONTROL

I think that a carefully thought-out battle strategy against pests, using both physical and chemical methods, is preferred. Systemic poisons placed in the potting soil are effective aginst insects that suck plant juices. The rationale for these poisons is that they are taken up from the soil by the plant's root system and transported throughout the plant by its vascular system. The poisons have no effect on the plant and pose no threat to pets or humans but have a deadly effect on the pests that suck the treated plant's juices. There are a variety of commercially available systemic poisons in the form of granular powders, liquids or impregnated sticks. These systemic poisons can be applied to the soil or misted onto the leaf surfaces, where they will be absorbed. Root systemic poisons have an added advantage in that they don't alter the glossy surface of decorative ornamental plants.

Spraying may be needed, but be sure to read the cautions on the label carefully and select the poison to match the pests and the plant. Some wide-spectrum insecticide sprays such as malathion can kill certain types of plants such as ferns or scheffleria. One strategy is to treat each plant individually by removing it from nearby plants, spraying it with the pesticide mist and then returning it. When spraying with either pressurized spray cans or hand-pump atomizers, make sure to wet both sides of the leaves and all of the stems and branches. If possible, it is best to spray the plant outdoors, where inhalation of the mist is much less than indoors and the smell of the spray is decreased.

Pesticides and their method of application are discussed elsewhere in this book (see page 144), and the pesticides of choice will be taken up in the text coverage of each type of insect pest. Remember: always read the label instructions and follow them carefully. Keep pesticide containers out of reach of children. When disposing of empty pesticide containers, follow the rules on page 179.

MEALYBUGS

Among the most insidious and persistent pests of household ornamental plants are the plant-juice-sucking mealy bugs. These multiply rapidly and can spread from plant to plant, seriously damaging or killing a great variety of indoor decorative trees, shrubs and plants. These bugs have piercing-sucking mouth parts; thus heavy infestations can cause severe loss of nutrients and result in wilting, dwarfing, leaf-drop and fruit-drop. These pestiferous insects also secrete large amounts of honeydew, a sugary secretion that provides a fertile growth medium for a variety of fungi, including black, sooty mold, which not only is ugly but also by coating leaf surfaces, prevents sunlight from reaching a plant's photosynthetic machinery. Honeydew also attracts ants that feed on this sticky-sweet nutrient. Ants literally farm both mealybugs and other honeydew-secreting insects such as aphids by carrying insects to new and fertile fields on other plants in the house.

Mealybugs, though small (an eighth to a quarter of an inch long), are easily detectable because they secrete a white, waxy, powdery material that coats their oval bodies with white, tufted, cotton-like projections. The infested plant looks as if it has dandruff. Mealybugs belong to the superfamily Coccoidea and the family Pseudococcidae. Males are rarely seen, and females, which are segmented and look like small woodlice, are hardly able to move. As adults they remain motionless, attached to the host plant

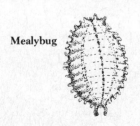

Mealybug

with their piercing/sucking mouth parts. Some mealy bugs reproduce parthenogenetically (without benefit of males), some bearing live young, others laying eggs. They are most often found where leaves and stems join.

In addition to the common mealybug, there is the citrus mealybug, which afflicts soft-stemmed and succulent ornamental houseplants, including jade trees, cacti, African violets, gardenias, fuschias and ferns. They are serious pests and often initially escape detection by hiding in crevices or other sheltered spots such as crotches where branches sprout from stems.

Control: Since mealybugs are often protected and transported by ants, an important first step is to control ants in the home (see page 119). Biological control by ladybug beetles will keep mealybug populations in check, since these colorful but voracious predators eat both adult and larval mealybugs. Some garden shops sell ladybug beetles and they survive and breed in greenhouses. Simple physical control with soapy water and washing with a strong hosing will remove most mealybugs (see page 143), and removing them mechanically with cotton swabs soaked in ordinary 75% rubbing alcohol is also effective. Physical methods combined with systemic poisons are often effective enough to eliminate the need of toxic sprays; however, should the infestation get out of hand, spray with insecticides that specify effectiveness against mealybugs. These include malathion, carbaryl, diazinon and acephate. We have had some success using a small soft paintbrush to apply the pesticide in any crevices where mealybugs thrive.

SCALE INSECTS

Closely related to the mealybugs are some rather atypical insects that are hardly recognizable as insects at all. These antennae-less, legless, motionless plant vermin are called scale insects. They do, however, have one very characteristic insect bit of anatomy—their piercing-sucking mouth parts that attach them to the infested plant parts. There are over 200 different kinds of scale insects. They all have a shell-like waxy armor or scale that protects the entire upper surface of the body. Most are only one-tenth to one-eighth of an inch in diameter, but some species may be as much as one-quarter of an inch in diameter. Their shapes vary, ranging from hemispherical to oval to oystershell-shaped. Their colors, depending on the species, range from white to black, with browns and grays being the predominant color pattern. Scale insects can be divided roughly into two groups, those whose scale is firmly and permanently attached and those whose scale can be easily removed.

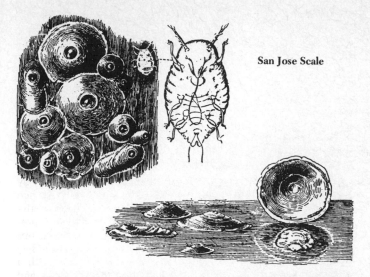

San Jose Scale

Both forms of scale insects are exceedingly prolific, producing whitish egg sacs or live young under the scale that eventually develop into tiny mobile nymphs that are responsible for the spread of the infestation. Unfortunately, scale insects are very inconspicuous and often are not detected until the host plant is seriously infested. Some forms of scale are found on leaves, others on both leaves and stems; still others attack only the stems and branches. They frequently show up in greenhouses and on decorative houseplants, such as fruit trees and other woody plants. They tend to form clusters, making the bark look as if it is coated with a flaky scale.

Control: If the infestation is only slight, physical methods can be effective. These include picking or scraping the scale insects off the infested plant and using cotton swabs soaked in rubbing alcohol or washing with soapy water. If some branches are heavily laden with scale insects, remove the branches and dispose of them. Dormant oil sprays, consisting of petroleum oil, fish oils or vegetable oils containing a toxin, are relatively safe and effective in treating scale insects. This strategy is effective because it not only introduces the toxin but also can mechanically block the insects' respiratory systems. Such dilute dormant sprays are best applied to stems and branches with a soft paintbrush but should not be sprayed on the foliage, as they can damage the leaves. Malathion, diazinon, carbaryl or methoxychlor added to the oil kills both the sessile adults and the mobile nymphs.

WHITEFLIES

A common pest of indoor plants is the easily recognizable but tiny whitefly. Whiteflies get their name from their covering of fine, white, waxy powder that gives plants infested with them the look of being covered in fine ash or tiny snowflakes. When such plants are moved, the winged adults take to the air, creating a cloud-like swirl of tiny white specks. These insects belong to the su-perfamily Aleyrodidae and resemble miniature moths whose wing-spans are only about one-fifteenth of an inch or less. They are quite prolific and produce large numbers of eggs that develop into flat, oval, pale green to whitish nymphs or crawlers. These larvae are initially active and then go into a motionless pupae-like state, losing their legs and antennae. During this period of their life cycle, they are protected by a wax-like collection of cast-off skin, but they continue to feed on the plant's juices until they metamorphose and emerge as adult flies. They are abundant in most areas but do best where it is warm.

These juice-sucking insects attach themselves to leaves and stems, most often on the undersurface of leaves. They digest plant sap and produce sugary, sticky honeydew, which then coats the leaves and promotes the growth of plant fungi, including the ugly, sooty black mold.

Whitefly infestations are easily detectible. They are usually brought into the house with newly purchased plants, particularly fuschias, lantana, tomato and many of the herbs. When buying these plants, check them carefully for signs of infestation. This will not be difficult, because whitefly infestations are easily de-tectable. The greenhouse whitefly, *Trialeurodes vaporariorum,* is also frequently found on azaleas, begonias, calceolarias, chry-santhemums, coleus, geraniums, hibiscus, roses and salvias.

Control: Spray plants, particularly the undersides of leaves, with spray combining malathion, rotenone and pyrethrins. Resmethrin

Whitefly

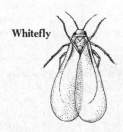

or acephate also have been reported effective against whiteflies. Wet the foliage and repeat applications three or four times at weekly intervals. Systemic preparations are also quite good at keeping whiteflies in check.

APHIDS, OR PLANT LICE

The superfamily Aphidoidea contains a vast number of families of plant pests that are the bane of every greenhouse owner and home gardener. These tiny (one-sixteenth- to one-eighth-inch-long), soft-bodied, rounded or pear-shaped, odious pests are called aphids or plant lice. Their bodies vary in color from green to pink to black. All have long, delicate legs and antennae, and in each species of aphid there are both winged and wingless forms, with the wingless being most numerous. Winged aphids at rest usually hold their wings up over the body like a peaked roof. Their abdomens are bulbous and their heads narrow and equipped with piercing-sucking mouth parts with which they suck plant sap. They secrete copious amounts of honeydew, a special, waxy, protective, fluid secretion, which may secondarily promote the growth of ugly plant fungi, blights, viruses and plant bacteria. Their saliva is believed to contain a toxic material that stunts plant growth and curls or otherwise distorts leaves. The sugar in their honeydew attracts ants that not only milk and nurture aphids but are known actually to farm these pests, carrying them to new and fertile pastures on uninfested plants.

Aphids have a prodigious reproductive potential, giving birth to live young without benefit of males. Winged forms occur whenever the infestation is heavy and uninfested plants are available. They are active from early spring through the summer. In the fall, sexual reproduction takes place as winged males and females mate; these now gravid females lay eggs that remain dormant throughout the winter. The combination of both sexual and asexual reproduction gives aphids a tremendous reproductive capacity, with a single female possibly producing trillions of eggs. Fortu-

Melon Aphid

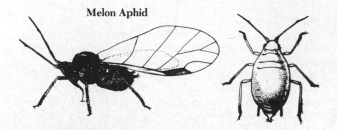

nately, this population explosion is usually reduced by a number of insects that feed on aphids. One such predator, the ladybug beetle and its larvae, consume large numbers of aphids. Larval green lacewings and larval syrphid flies also are mighty aphid eaters.

Aphids attack most plant parts, concentrating their efforts on succulent, tender young leaves, stems and flower buds. Other aphids carried by ants feed only on plant roots and hence are not often seen. The visible forms can be seen in massed clusters, and their secretions impart a shiny appearance to plant foliage.

Control: First, wash off the plant with soapy water and spray with a strong stream of water. This will get rid of a great many of the pests. If you can purchase ladybugs in your area, they will really keep the indoor aphid population down. Since ants are often co-conspirators with aphids in the assault on your ornamental plants, it would be wise to control the ants in your home (see page 119). If the infestation is in its early stages, systemic pesticides can be quite effective, but if the infestation is severe, spraying will be needed. Use commercially available sprays or dusts, making sure that the label specifies "Aphids" and that the product may be used on houseplants. Malathion, rotenone and nicotine sulfate are effective aphicides, but treatment must continue weekly through the spring and summer or the aphid population will explode.

CUTWORMS (CATERPILLARS)

There are several species of the order Lepidoptera whose larvae can be troublesome pests in greenhouses. These so-called cutworms may range in size from the barely visible to up to two inches in length when full grown. Some are solidly colored, some striped lengthwise, some mottled and some ringed. Colors are usually shades of brown or green or combinations of brown, red, yellow, green, gray and black. Some are smooth-skinned and some are "hairy." Some species hide deep in the soil or deep in flowers and are difficult to detect, while others feed above ground and can be found anywhere on the plant. Their preferred foods are succulent leaves, stems and buds. These pesky caterpillars develop from eggs laid by night-flying moths that get into open ventilators or through unscreened windows or are brought into the house with newly purchased plants.

The damage can be severe, with leaves and buds partly or entirely eaten. Sometimes whole branches or stems are cut off, often near the soil. These voracious eaters will often leave pellets of dark excrement on the plant.

Control: Hand picking of larger cutworms is an effective form of control. For smaller cutworms and soil-dwelling cutworms, use a spray containing diazinon, making sure that on the label it lists "caterpillars" as one of the insects controlled. As usual, follow instructions exactly. Since these pests are hard to control, you may have to repeat at weekly intervals.

EARWIGS

There are some thousand species of the insect order Dermaptera that are commonly called earwigs. For many people they are notorious, loathsome creatures provoking shudders because of the ancient myth that they crawl into the ears of sleeping humans, cut their way through the eardrum and lay their eggs in the brain. This theme was utilized in the second *Star Trek* movie, in which a mythical earwig-like creature does just that. The myth may have had some basis in fact—insects have been known to crawl into ears—but by and large this happenstance has about the same probability as winning the Irish Sweepstakes. Three or so species of earwigs can be house pests, damaging plants and generally making homeowners anxious. These include the ⅝-inch-long European earwig, the inch-long ring-tailed earwig and the shore or striped earwig, an inch-long inhabitant of tropical and subtropical areas.

Earwigs are generally brown to blackish in color; they have broad heads bearing large, compound eyes and long, slender, multi-segmented antennae. They are elongate insects with short but broad front wings and are equipped with a rather remarkable armament, a strong, curved set of pincers that project out behind the animal. The forceps-like cerci may be from ⅜ to a quarter-inch in length and in the males are strongly curved. These pincers

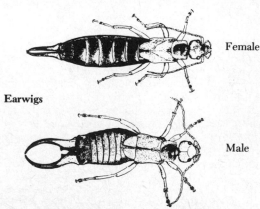

Female

Earwigs

Male

are mainly defensive weapons and occasionally can be used to cause humans a painful, bruising but harmless pinch.

Earwigs are cosmopolitan and found throughout the United States and southern Canada. The female lays eggs and overwinters with them; unusual among insects, the female earwig is a good and caring mother. The eggs hatch and the nymphs remain with the family unit until mature enough to wander off on their own.

They are minor plant pests, damaging flowers and leaves when they emerge from their daytime hiding places. They like dark, moist areas but are frequently found under rugs, in laundry left in the basement and in cushions of porch furniture left outdoors overnight. They can enter the home from debris piled outside or may be brought in with cut flowers or firewood. Some earwigs are equipped with scent glands and can give areas of heavy infestation a foul odor. In the warmer months, they may swarm at night and invade homes in considerable numbers.

Control: Remove earwig hiding-places around the outside (boards, piles of debris, mulch, etc.) and spray a five-foot-wide perimeter around the house with bendiocarb, diazinon or carbaryl sprays. Treat crawl-spaces with dusts and granules. For quick knockdown, pyrethrin-based insecticide sprays are effective. Poisoned baits containing 2% propoxur are also effective when dispensed according to label instructions.

THRIPS

Rarely noticed but present in large numbers, particularly on flowers, are the slender, almost microscopic (1/20- to 1/16-inch-long) members of the insect order Thysanoptera. These tan/brown/ blackish, tiny insects have two pairs of very narrow, hairy wings, hence are sometimes called "bristlewings." Their young are wingless, yellow to orange and are often splattered with black dots of thrips excrement. Both young and adults have piercing mouth parts with which they suck plant juices and produce considerable damage to flowers and leaves. Evidence that thrips have infested your houseplants is provided by silvery, irregular streaked areas specked with tiny black dots. If a thrips-infested plant is disturbed, hordes of adults may take to the air and may get into your eyes. Some people call thrips thunderflies, since in the wild they rise in dense swarms when disturbed by thunder. They have a considerable reproductive capacity and, given their two- to three-week life cycle, can become an episodic major pest.

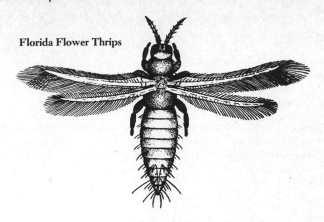

Florida Flower Thrips

Control: Spray with malathion-containing houseplant sprays at two-week intervals. Systemic pesticides may be of considerable value, since thrips are plant-juice eaters.

SPRINGTAILS

These distinctive insects of the order Collembola derive their name from a remarkable, forked, muscular structure at their rear ends. By jerking this organ they can propel themselves with some spectacular flips when disturbed. They, like all insects, have six legs and can also move about by more conventional walking. They are wingless, very small (less than one-fifth of an inch in length), slender and segmented or globular in shape. They vary considerably in color (whitish/purple-gray/black). There are 500 species of springtails, about 20 of which can be troublesome pests in the soil of flowerpots and greenhouses, where they chew seedlings and tender plant parts. They also may infest basements, kitchens and bathrooms, sometimes in considerable numbers, and are often found in light fixtures, since they are attracted to lights.

They can survive very cold temperatures and, when environmental conditions improve, can reproduce quite quickly. They do best in high humidity or in moist areas such as drains, damp basements or bathrooms, where they hide in cracks that have accumulated organic debris. They are often found on moist soil surfaces of houseplants.

Control: A variety of commercially available sprays (make sure the label says "house plants" if spraying plants) containing either

2% malathion, 0.5% diazinon or 0.25% pyrethrins are effective, but treatment should be repeated in about three weeks. Spray pots, soil, the plants themselves and shelves. In the home, remove any debris and spray around cracks and fissures where the forked-tailed little pests hide out.

SLUGS AND SNAILS

Slugs and snails *(escargots)* in some places are sought-after gourmet delights, served with herbed garlic butter, but to most homeowners these squishy, slimy mollusks can be a major problem to potted plants indoors and in greenhouses. I've had my entire crop of lettuce seedlings devoured by these pests and, indeed, they can make a meal of all kinds of leafy plants, particularly on those with leaves on or near the soil. They are land-dwelling members of the phylum Mollusca, class gastropoda, which means "stomach-foot." They move by means of a muscular foot and secrete copious amounts of mucus as they glide along. Their proboscides contain an extrusible, rasping band-saw-like organ, with which they macerate tender plant parts before ingestion. The first evidences of their presence are the shiny mucus trails they leave behind and the round holes they chew in leaves.

There are 30 or so species of slugs and snails that are invaders of our homes. They range in size from a quarter-inch to ten inches in length, are nocturnal and may live for several years. They avoid dusty, dry places and are found under pots, in and under decaying wood, in moist organic debris and rockpiles. They are most active after rains or on moist, foggy nights. They lay eggs (ten to 200) in moist places from spring to fall; these develop into adults in about a year. If conditions are unfavorable, they can go into a dormant state emerging again when environmental conditions are more favorable. Their colors range from gray to brown and some are mottled. Slugs have no shells, but snails have a coiled shell—indeed, some tropical land snails have spectacularly beautiful shells. One recent pest introduced into southern Florida is the

Snail

Slug

large African land snail that devastated much of the vegetation on the Islands of Guam and Okinawa when it was introduced there as a food source by the Japanese.

Prevention and Control: Remove their usual hiding and breeding places. They can't travel very far; hence, once you've cleaned up debris around the house (e.g. sunken door-wells and window-wells) and in the basement or greenhouse, you will be rid of them. Be particularly careful with potted plants put outdoors for the summer. I've found the pot bottoms literally crawling with slugs. Slugs and snails love beer, even stale, flat beer, and they can be trapped and drowned in dishes of beer put out in your greenhouse or garden. There are also commercially available poison baits having such descriptive names as Slugbegone. These baits contain 2% methiocarb. Scatter moist mounds of bait around areas where you see slug trails or plant damage. Cover the bait with a propped-up flat rock to prevent your pets or children from getting to the poison. Baits should be put out several times at three-week intervals.

SOWBUGS AND PILLBUGS

These occasional troublemakers in home greenhouses are arthropods that belong to the class Crustacea, thus are more closely related to crabs, shrimp and lobsters than they are to insects and arachnids. As a matter of fact, they are among the very few crustaceans that have adapted to living on land, but still breathe by

Greenhouse Pillbug

means of gills, which means they must have a moist environment. Sowbugs and pillbugs look like miniature armadillos. Usually, they are less than three-quarters of an inch in length and have oval bodies that are rounded above and flattened below. They are covered with seven overlapping external armored plates and move about on several pairs of legs. Their small heads project out from the armor and bear a pair of antennae.

The common pillbug *(Armadillidium vulgare)* is a frequent inhabitant of greenhouses, damp basements and houses built on slabs. It is found, sometimes in large numbers, under moist organic debris and in the soil under boards, woodpiles and even in moist, decaying beams that have been damaged by termites or carpenter ants. They can do considerable damage to seedlings and roots or tender new sprouts of houseplants. Pillbugs when disturbed tend to roll up into a tight ball.

The garden sowbug, *Porcellio laevis*, is similar in size and shape to the pillbug but has two long, pronged tail appendages that protrude from the rear of the bug's plated armor covering. Sowbugs, unlike pillbugs, can't roll into a ball. Both of these pests produce two or three generations a year. They are active at night during the warmer months.

Prevention and Control: First, remove potential nesting and hiding areas from greenhouses, basements and around the outside of the house. If there is a major sowbug problem, spray a band around the building, particularly around doorways and window-wells, with: diazinon (four fluid oz./gal.) or bendiocarb (one packet/gal. H_2O) or propoxur powder (two oz./gal.). Do not get the last-named chemical on your plants: it could damage them.

PLANT MITES

Plant mites are not insects but are arachnids directly related to spiders, chiggers, scorpions and ticks. Like all arachnids, plant mites have eight legs. While not all mites are detrimental, those that feed on plant juices are insidious, destructive pests that severely damage plants. The most prevalent and widespread mite is the minute red spider mite (also called the two-spotted spider mite or false spider mite). These pests are found throughout the United States and southern Canada, particularly in the warm, dry days of midsummer.

These flat, oval, dark red to green or yellow-brown mites are barely visible to the naked eye. Adult mites might be detected by shaking a branch over a sheet of white paper: if the specks on the paper move, you've got a problem. Actually, these almost

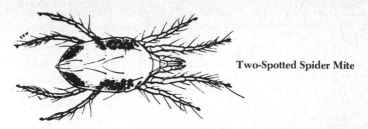

Two-Spotted Spider Mite

invisible mites do announce their presence by their delicate webs holding bright red eggs that can be seen only under a microscope. There are usually plenty of webs, since their life cycle is short, and every week or ten days a new generation is produced, mostly on the undersides of leaves along the leaf veins or other irregularities on the leaves.

The damage the mites do is also a giveaway. Look for finely stippled, bronze or rusty-brown areas along the veins on the undersides of leaves. Edges of infested leaves deprived of nutrients may die or the leaves may lose color and drop off. In severe infestations, the weakened plants become prey to disease and may die.

Another common ornamental plant mite is the cyclamen mite, which frequently attacks African violets, cyclamens and other houseplants. The adult mites are too small to be seen with the naked eye, but under the microscope or a good magnifying glass they are visible as oval, semitransparent amber or tan-colored, glistening, eight-legged mites. Their round, milky white eggs look like miniature pearls under the microscope.

The mites usually seek protected feeding areas on young, tender leaves or stem ends, buds and flowers. The mites can move, albeit slowly, and they can crawl from leaf to leaf whenever there is contact. They can also be spread to healthy plants by contact with hands or clothing that has touched an infested plant.

Evidence of infestation is derived mainly from the type of damage these pests wreak on their victim. Leaves will twist and curl and become brittle. Buds will form but fail to open, and the flowers that do develop are often deformed and streaked. Blackening of cyclamen mite-injured leaves, buds and flowers is also a common symptom.

Control: If mite infestations are heavy, the best advice is to get rid of the diseased plant. Put it in a plastic bag, seal it tightly and put it out in the trash. Make sure to wash off hands and clothing that might have come in contact with the infested plant. Since the mites thrive in dry, warm weather, you can deprive them of their best environment by watering frequently during the dry midsummertime.

Mite-infested plants should be washed with warm soapy water. Cyclamen mites can be killed by immersing the whole infested plant in hot (110°F) soapy water for ten minutes. Higher temperatures are even more effective, but they also will probably kill the plant, so be sure to use a thermometer when trying this tactic.

MILLIPEDES

Millipedes, arthropods of the class Diplopoda, are usually found only outdoors but occasionally invade households and may do damage in greenhouses by eating bulbs, tender roots and leaves. The diplopoda all have cylindrical, segmented bodies with two pairs of legs on each segment—except for the first three segments, which have only one pair of legs apiece. There are more than a thousand species in the United States and Canada, but all look pretty much alike. They are wormlike in appearance, are about an inch and a half long as adults, are usually brown, tan or gray and are covered with a hard, armor-like external skeleton. Their many short legs allow them to move along at a slow but steady pace. They are commonly called thousand-leggers and usually come out only at night. If disturbed, they tend to curl into a circle. They protect themselves with a malodorous secretion and in heavy infestations can produce a rather unpleasant smell. For reasons unknown, millipedes sometimes invade houses in very large numbers.

Millipedes favor moist, sheltered areas, such as piles of organic debris. They can be found under boards and under flowerpots. They are often brought into the home with potted plants which have been put outdoors during the summer.

Prevention and Control: Millipedes will usually die in a dry house; thus, the best way to control them is to remove their hiding places. Remove any damp, decaying organic matter on your greenhouse floor or in and around your basement and remove any buried

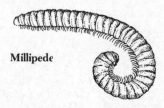

Millipede

wood or loose wood near the house. Should infestations become heavy, block entrance to your home by spraying a five-to ten-foot-wide band around possible entry sites, door frames, windows and window-wells, with copious amounts of malathion, Baygon, Ficam or Sevin.

PART VII
Lawn Pests

Many insects and insect-like pests damage lawns and other turf. They cause the grass to turn brown and die or they build unsightly mounds that may smother the grass.

Some of the pests infest the soil and attack the plant roots; some feed on the plants' leaves and stems; others suck juice from the plants.

Other insects and insect-like pests inhabit lawns but do not damage them. The pests are annoying, and some of them attack people.

These pests can be controlled with insecticides. The recommendations here are applicable to lawns, which are extensions of our homes. However, they are *not* intended for the control of insects on turf areas that might be grazed by livestock.

LAWN INSECTS: HOW TO CONTROL THEM

PESTS THAT INFECT SOIL AND ROOTS

Grubs

Grubs are the larvae of several species of beetles. They are whitish or grayish, have brownish heads and brownish or blackish hind parts, and usually are found in a curled position when disturbed. They hatch from eggs laid in the ground by the female beetles. Most of them spend about ten months of the year in the ground; some remain in the soil two or three years. In mild weather, they live one to three inches below the surface of the lawn; in winter they go deeper into the soil.

They burrow around the roots of the grass and feed on them about an inch below the surface of the soil. Moles, skunks and birds feed on the grubs and may tear up the sod in searching for them.

This section is based on Home and Garden Bulletin No. 53 of the U. S. Department of Agriculture.

You can estimate the grub population of your lawn. Do this in the fall before cold weather sets in or in the spring after the soil warms up and the grubs are near the surface. With a spade, cut three sides of a strip one foot square by two or three inches thick. Force the spade under the sod and lay it back, using the uncut side as a hinge. Use a trowel to dislodge soil on the overturned roots that might contain grubs. Count the grubs in the exposed soil. Replace the strip of sod. In the same way, cut strips of sod in several other parts of the lawn and count the grubs under each strip. To calculate the average number of grubs per square foot of lawn, divide the total number of grubs counted by the number of strips. If the average number is three or more, apply an insecticide.

The parent beetles differ in appearance, distribution and habits. The following are important pests of lawns:

May Beetles: These beetles are brown or blackish brown. More than 200 kinds are found in the United States. Sometimes they are called June beetles.

In Indiana and southern Wisconsin, the beetles are found from the first of May to mid-July; they are most abundant during the last two weeks of May. South of this area the beetles appear in early April; north of it they appear in late May and are most numerous in June. In Arizona, most kinds are found during July and August.

The young are called white grubs. Some of them remain in the soil two or three years and may feed on the grass roots during several seasons.

Japanese Beetle: This beetle is about half an inch long and has a shiny metallic-green body; it has coppery-brown wings and six small patches of white hairs along each side and the back of the body, just under the edges of the wings. The adult insect feeds on many different plants.

Japanese beetles are found mostly in the eastern states. They appear about May 15 in eastern North Carolina, June 15 in eastern Pennsylvania and July 1 or later in New England. They are active

**Japanese Beetle
and Grub**

for from four to six weeks. In North Carolina, most of them die early in August; in New England some are present until frost.

Asiatic Garden Beetle: The beetle is about a quarter-inch long, is chestnut brown and has a velvety appearance. The underside of the body is covered with short yellow hairs. The insect flies only at night and feeds on various kinds of foliage.

These beetles are found in widely scattered places along the Atlantic seaboard from Massachusetts and eastern New York to South Carolina. They are most abundant from mid-July to mid-August.

Oriental Beetle: This beetle is about ⅝ of an inch long, is straw-colored and has some dark markings on the body.

The beetles occur in Connecticut, Massachusetts, southeastern New York and northern New Jersey. They appear in late June, July and August. The grubs prefer unshaded lawn and short grass.

European Chafer: The beetle is about half an inch long and is light chocolate brown or tan. The insects emerge from the soil at dusk, swarm into the trees and shrubs and make a buzzing sound.

European chafers occur mostly in New York; isolated infestations have been found in Connecticut and Virginia. The beetles appear in June and July and are most abundant in early July.

Masked Chafers: These beetles are half an inch long and brown. They live in the soil during the day and emerge at night; they are especially active on warm humid evenings.

The northern masked chafer is found from Connecticut south to Alabama and west to California. The southern masked chafer is common in the southeastern states; it is found throughout the south and in Texas, Oklahoma, Iowa and Illinois. Masked chafers appear in late June and July and are active one or two months.

The young are sometimes called annual white grubs because the life cycle of the insect is completed in one year. The grubs have irregularly arranged spines on the underside of the last body segment.

Rose Chafer: The beetle is half an inch long and is yellowish brown; it has long, spiny legs. Rose chafers feed on almost any vegetation and are very destructive to roses in bloom. They prefer areas in which the soil is light and sandy.

The insects are found in the eastern United States and west to Colorado and Texas. They are abundant in June and early July.

The grubs are not so harmful to lawns as some of those mentioned above.

Green June Beetle: The beetle is nearly one inch long. The body is somewhat flattened; it is velvety green and has bronze to yellow edges. The insects feed on the foliage of many trees and plants. The females often lay eggs in piles of grass clippings, as well as in soil.

Green June beetles are found mostly in the southern part of the United States but frequently as far north as Long Island and southern Illinois. They are active in June, July and August, and produce one generation a year.

The grubs feed mainly on decaying vegetable matter. Their burrowing mounds the soil and smothers the grass. The grubs also uproot seedlings in newly sown lawns. Damage is most severe in dry seasons and is most apparent in the fall.

Sometimes after a heavy rain, the grubs come out of the soil and crawl on the surface of the ground. They have the unusual habit of crawling on their backs.

MOLE CRICKETS

Mole crickets are light brown; their lower surface is lighter than the upper and is often tinged with green. They are about one and a half inches long and have short, stout forelegs, shovel-like feet and large, beady eyes.

Mole crickets feed on the roots of the grass. In addition, their burrowing uproots seedlings and causes the soil to dry out quickly. One mole cricket can damage several yards of a newly seeded lawn in a single night.

These insects are most numer us in the South Atlantic and Gulf Coast states from North Carolina to Texas.

WIREWORMS

Wireworms, the larvae of click beetles, are from half an inch to one and a half inches long and are usually hard, dark brown, smooth and slender. Some wireworms are soft and white or yellowish. Wireworms bore into the underground part of the stems and feed on the roots of the grass. The boring causes the plant to wither and die.

The adults are about half an inch long and brownish, grayish or nearly black. They are hard-shelled, and their bodies taper somewhat at each end. When they fall on their backs, the beetles, in an attempt at righting themselves, flip the middle part of the body against the ground, throw themselves several inches into the air, producing a clicking sound.

PERIODICAL CICADA

The young, or nymphs, leave many small holes in lawns, especially under trees, when they emerge to become adults.

If you hear the day-song of the cicada in the spring of a year in which a brood is scheduled to appear in your region, the holes in your lawn were probably made by the emerging nymphs.

If a large brood is emerging, control of adults or protection of ornamental trees and shrubs in the lawn is recommended.

BILLBUGS

The young, or grubs, are small and white and have hard brown or yellow heads. They feed on the roots of the grass.

Adult billbugs are beetles a fifth to three-quarters of an inch long. They have long snouts or bills that carry at the tip a pair of strong jaws with which the beetles chew their food. Their color is clay yellow to reddish-brown to jet black. The beetles burrow in the grass stems near the surface of the soil and also feed on the leaves.

Several species of billbugs damage lawns. One species, *Sphenophorus phoeniciensis*, attacks Bermuda grass in Arizona, California and New Mexico. Another species, *S. cicatristriatus*, has damaged lawns in eastern Washington. The hunting billbug has caused extensive damage to zoysia grass from Maryland to Florida and in California and Hawaii.

PESTS THAT FEED ON GRASS LEAVES AND STEMS

Sod Webworms

Sod webworms are about three-quarters of an inch long and light brown. Their bodies are covered with fine hairs.

Sod webworms are the larvae of lawn moths. The adults are small, whitish or gray moths (or millers). They fold their wings closely about their bodies when at rest. They hide in shrubbery or other sheltered spots during the day. In the early evening they fly over the grass and the females scatter eggs over the lawns.

The worms work only at night. They live in protecting silken webs or nets that they form about their bodies. As soon as they are hatched, they start feeding on blades of grass. When they grow larger, they build burrows or tunnels close to the surface of

Sod Webworm

the soil; they reinforce the tunnels with bits of dirt and pieces of grass, line them with silk and live in them. They cut off blades of grass and eat them. Some species feed on the grass crowns at ground level and on the roots. As partly grown larvae, they overwinter in their silken webs.

Sod webworms prefer new lawns. They attack bentgrass, bluegrass, fescue and other grasses. Irregular brown spots are the first signs of damage. If the infestation is heavy, large areas of grass may be damaged severely or destroyed in only a few days.

Several species infest lawns. The tropical sod webworm is the most important one in Florida. A burrowing sod webworm sometimes infests lawn grasses from Kansas south to Louisiana and east to Maryland. The dirty white larvae live in silk-lined tubes about ⅜ of an inch in diameter that extend two to three inches into the soil.

You can find the worms by breaking apart some of the drying sod. If there are three or four of them within a six-inch-square section, apply an insecticide.

Armyworms

Armyworms are the larvae of moths. They are one and a half inches long; they are greenish and have blackish stripes along each side and down the center of the back.

The adults are brownish gray; their wings measure about one and a half inches across when expanded.

Armyworm

The armyworm and the fall armyworm are common species. When they are numerous, they may devour the grass down to the ground. Their feeding causes circular bare areas in lawns. The lawn armyworm sometimes damages lawns in Hawaii.

Cutworms

Cutworms are dull brown, gray or nearly black caterpillars and are one and a half to two inches long. Some cutworms are spotted;

others are striped. Usually they hide in the soil during the day and feed at night. They are the larvae of night-flying brown or grayish moths.

Cutworms occasionally infest lawns. They feed on the leaves or cut off the grass near the soil and may do severe damage to seedlings of Bermuda grass, bentgrass and rye grass.

Fiery Skipper

The larvae of the fiery skipper feed on the leaves of common lawn grasses but attack bentgrass most severely. Early infestation is indicated by isolated, round, bare spots one to two inches in diameter. The spots may become numerous enough to destroy most of the grass on the lawn.

The adults are small, yellowish-brown butterflies.

The fiery skipper is occasionally a pest of lawns in California.

Lucerne Moth

The larvae of this insect prefer clover and other legumes, but they also infest grass.

The adult is a grayish-brown moth; it has two pairs of dark spots on each forewing.

The lucerne moth sometimes attacks lawns throughout California. It has been recorded in 20 other states across the country.

Leaf Bug

The leaf bug feeds on lawn grass and causes it to die out in spots. This insect is gray and white. It attacks bluegrass, Bermuda grass and bentgrass in California and is distributed throughout the United States.

Pests That Suck Grass Juices

Chinch Bugs

Most chinch bug damage is caused by the young bugs, or nymphs. Yellowish spots appear in the infested lawn; they turn rapidly into brown, dead areas.

Nymphs hatch from eggs laid by the female adults. At first a nymph is about half the size of a pinhead; it is bright red and has a white band across the back. As it grows, it sheds its skin four times. The full-grown nymph is black and has a white spot on the back between the wing pads.

The adults are about a sixth of an inch long; they are black and have white markings.

Chinch Bugs

The species *Blissus leucopterus hirtus,* the hairy chinch bug, infests lawns in the eastern part of the United States. The species *B. leucopterus leucopterus* is a pest of Bermuda grass in Oklahoma. *B. insularis,* the southern chinch bug, severely damages St. Augustine grass in Florida and other Gulf States. It also attacks other grasses in these states and is a serious lawn pest in Georgia, North Carolina and South Carolina.

In the east and as far south as northern Florida, the eggs hatch in the spring and nymphs infest lawns until late fall. The adults hibernate during the winter. In southern Florida, they are active during the winter, except on the coldest days.

Scale Insects

Scale insects suck the juice from grasses—some feed on the crown plants and aboveground parts; others feed on the roots. The grass becomes yellow and then brown and finally dies. Damage is usually more severe in dry periods than in wet.

Several kinds of scales damage lawns in the southern part of the United States from South Carolina south and west to California. No satisfactory way has been found to control scales on lawns. Consult your county agricultural agent or state agricultural station for current recommendations.

The most important species of scales are the following:

Rhodes Grass Scale: The adult is about an eighth of an inch in diameter, globular, dark purplish brown and covered with a white cottony secretion. The tiny nymphs, or crawlers, move about at first and then settle down to feed. They secrete a wax that covers them. The Rhodes grass scale may produce five generations a year.

The adults and nymphs cause lawn damage. They attack chiefly the plant crowns and cause infested plants to turn brown and die.

The Rhodes grass scale is found in Arizona, California, Hawaii,

New Mexico and the Gulf States. It attacks Bermuda grass and St. Augustine grass.

Bermuda Grass Scale: The adults are about a sixteenth of an inch long, oval and covered with a white, hard secretion.

This insect infests Bermuda grass and is especially active in shady areas. It kills the grass and leaves bare brown patches.

Ground Pearls: The female adult secretes a white wavy sac in which it places about a hundred pinkish-white eggs. Slender nymphs hatch and feed on the fine grass rootlets. The nymphs cover themselves with hard globular shells that look like tiny pearls. These are called ground pearls. They are about an eighth of an inch in diameter.

Ground pearls cause serious damage to Bermuda grass in the South and Southwest and to centipede grass in the South. The grass that they attack turns brown in the summer; it dies in the fall and leaves irregular dead spots.

LEAFHOPPERS

Leafhoppers are tiny triangular or wedge-shaped insects that fly or hop short distances. They are less than a fifth of an inch long and green, yellow or brownish-gray.

Many species of leafhoppers infest lawns. They suck the sap from the leaves and stems of the grass. New lawns may be damaged so extensively that reseeding is necessary. Damage to established lawns is evident in whitened patches. It is often mistaken for damage due to dry weather or disease.

MITES

Several species of mites attack grasses. They suck the sap and cause the leaves to be blotched and stippled. Severe infestations can kill the plants.

The banks grass mite occurs throughout most of the United States and occasionally attacks lawns. It is not ordinarily a pest in well managed lawns.

The Bermuda grass mite damages Bermuda grass throughout the extreme southern United States. It is very tiny and light beige in color. It causes shortened stems and rosetted growth.

Another mite, *Oligonychus stickneyi,* damages Bermuda grass in New Mexico, Arizona and California. In Florida, it damages other grasses as well.

Clover mites feed on clover and other lawn plants. They are very tiny and red. Although they feed only on plants, they are a nuisance when they enter homes. This usually occurs in spring and fall.

CONTROL

The pesticides mentioned in the following section are available in several different formulations that contain varying amounts of active ingredients.

As usual, users are cautioned to read and follow all directions and precautions given on the label of the pesticide formulation that will be used.

Insecticides are sold under various trade names by garden supply houses and hardware, seed and drug stores.

Granules are ready-made formulations that are used dry. Apply them with a lawn fertilizer spreader.

Wettable powders and other formulations are used in sprays. Mix the purchased product with water and apply with a garden-type compressed-air sprayer or a knapsack sprayer. The quantity of water to use depends on the type of sprayer you have. If a wettable powder is used, frequent agitation of the mixture is necessary.

A quart-jar attachment for a garden hose will provide good distribution of an insecticide on a lawn. Use an attachment that delivers a coarse spray and a large volume of water. Usually a quart-jar full of an insecticide mixture will cover about 500 square feet of lawn.

Baits are usually purchased ready mixed, but a bait for controlling slugs and snails may be prepared.

Control of soil insects is sometimes difficult. It is therefore important to apply the pesticide at the time of the year when the insect is most susceptible to control. This information will be on the pesticide label.

To control underground lawn pests, apply an insecticide and, immediately afterward, sprinkle the lawn thoroughly. One application may control the pests for several years. Control of soil insects is slow. It may be a month or more before the insecticide becomes fully effective.

To control aboveground lawn pests, apply an insecticide to the grass. Sprinkle lightly with water to wash the insecticide down around the crowns of the plants. Do not water again for a few days; then sprinkle the grass thoroughly to wash off the insecticide. One application may control the pests for several weeks. Repeat the application if they become numerous.

Table 1 gives insecticides recommended for most lawn pests.

Table 1

LAWN PESTS AND INSECTICIDES TO USE IN THEIR CONTROL[1]

Lawn pest	Diazinon (Spectra-cide)	Carbaryl (Sevin)	Chloropyrifos (Dursban)	Trichlorofon (Dylox)
Ants[2]	X	X	X	X
Armyworms	X	X	X	X
Billbugs	X	X		
Chiggers	X		X	
Chinch bugs	X	X	X	
Cicada killer wasps	X			
Cutworms[3]	X	X	X	X
Earwigs	X	X	X	
Fleas	X	X	X	
Fruit flies	X			
Grasshoppers			X	
Grubs[4]	X	X	X	X
Leafhoppers	X	X		
Millipedes	X	X		
Mites, clover	X		X	
Sod Webworms	X	X	X	X

[1]Several insects are not listed, either because no control measures are necessary or chemicals for their control are not registered at this time.

[2]If only a few ant nests are present, treat them individually. Wash the insecticide into the nests or drench the mounds with it. Special treatment is required to control fire and harvester ants; consult your state's Agricultural Experiment Station for latest recommendations.

[3]To control cutworms, apply the insecticide in late afternoon.

[4]In hot, dry areas, lower dosages may be necessary to prevent burning the grass; consult your state's Agricultural Experiment Station.

MILKY DISEASE

Grubs of the Japanese beetle can be controlled by applying a dust containing spores of milky disease. However, the disease may require several years to spread and reduce beetle populations appreciably.

In general, the spore dust should not be applied to soil that has been treated with an insecticide. Although the insecticide does not harm the disease spores, it reduces the grub population and thereby greatly lowers the chance of establishment and spread of the disease.

Spore dust powder can be purchased. Apply it when the ground is not frozen. Follow directions on the label.

PART VIII

Chemical Control of Pests and Its Dangers

Throughout our lives, we are presented with a series of risk-benefit decisions. One of these decisions, faced by homeowners having problems with an invading pest, is whether to use some chemical pesticide to get rid of that pest, balanced by the risk of that substance harming humans, pets and harmless wildlife. Obviously, such a decision requires some understanding of what these chemicals do and what amount of the chemical is dangerous. Unfortunately, there is much anxiety about the use of poisons and the real risk as opposed to the imagined risk. In the following sections, several relevant topics are discussed that should help you make an intelligent risk-benefit decision, including guidelines for the storage and disposal of insecticides, how to prevent accidental poisoning and what to do should accidental poisoning be suspected.

ACUTE PESTICIDE POISONING

Because of the highly emotional public debate over the hazards of pesticide poisoning, there has been a tendency to distort the problem of acute pesticide poisoning. "Acute" refers to poisoning that causes a rapid onset of illness or death; "chronic" refers to long-delayed effects. While the exact numbers of acute pesticide poisoning cases are difficult to acquire, we can come up with a reasonable estimate of frequency of intoxication. Deaths from all forms of acute poisoning—plant, animal, medications, industrial chemicals and pesticides—occur at the rate of about two per hundred thousand population per year. Pesticide poisoning is responsible for less than 1 percent of lethal acute poisoning. Thus, the acute death rate for pesticide poisoning is less than one per million population. The number may even be on the decline, since many of the more toxic pesticides have now been banned. These are pretty good odds; still, more than one hundred deaths per year are caused by acute pesticide poisoning. Deaths alone are not the whole story: there may be 25 to 100 times as many cases of pesticide poisoning that cause either temporary symptoms or illness.

The greatest incidence of acute poisoning occurs in children under ten. Some ingestions of poisons by adults showed that these were deliberate (i.e., suicide attempts); others were by inebriated or senile individuals or illiterates who couldn't read labels. Thus, most of the deaths associated with acute pesticide poisoning were purely accidental and preventable.

CHRONIC PESTICIDE POISONING

The greatest concern of the public and of the Environmental Protection Agency is the chronic (continued) exposure of the population to long-lived pesticides that accumulate in the environment. Such relatively non-biodegradable chemicals may enter our water supplies, accumulate in our crops or be incorporated into animal and human tissues such as body fat. As exposure continues, there is fear that concentrations may reach levels that may cause a variety of long-term effects, including genetic damage, an increased frequency of some cancers, birth defects, brain damage, liver damage and kidney damage. Studies of permissible human pesticide levels for now-banned insecticides such as dieldrin and DDT were on the order of one part per million. However, the question of how much is too much is very difficult to determine, particularly since there is some excretion of the toxin and recovery from chemically caused damage. In general, the maximal permitted doses and rates of exposure to these toxins is 100 to 1,000 times less than that which is known to cause injury or chronic illness. The stringent guidelines for exposure defined by various regulatory agencies reflects our concern for possible undiscovered effects. While there is no reason for undue anxiety if one follows the instructions on the label, it is worth noting that any potentially toxic compound should be treated with caution and every attempt should be made to minimize exposure even to amounts known to be harmless.

PREVENTION OF POISONING

Every year, 850,000 children under the age of five are exposed to poisons and reported to poison control centers. This means that five out of every hundred children experience such an event, but most, 90 percent, are not seriously affected, and fewer than 1 percent of the cases result in death. Most of these poisonings occur in the home and are preventable, mainly by denying children access to toxic substances. Access, however, is not the only factor to be considered. Stresses in the home environment are often associated with accidental poisoning of children such as moving to a new home, illness or divorce in the family, pregnancy, etc.). Parents under stress are not alert and consequently may not take the necessary precautions to prevent accidental poisonings. Most victims are between 18 months and three years of age. Boys are at higher risk than girls.

Education is one key to prevention. Children should be taught to recognize labels and parents should read labels carefully. The enactment of the Poison Prevention Packaging Act in 1970 has helped reduce emergency room visits for poison ingestion by almost 50 percent in the last five years. However, real prevention begins with you, the homeowner. There are some simple guidelines that will help you prevent accidental poisonings. Make sure that pesticides are stored in locked cabinets and are out only when in use. Be careful when disposing of pesticide containers, since they may have enough toxin left to poison a child. Write your local poison control system telephone number in the next section in the place marked and read this section carefully. If you employ baby sitters, make sure that they are aware of what this section advises.

ALWAYS READ THE LABELS ON POISON CONTAINERS

Every label should be read carefully and the instructions on the label followed exactly. Each label has an Environmental Protection Agency (EPA) number, which is your indication that this chemical is effective and without hazard when you follow application instructions. Some people feel that if 0.5% is effective, then 1.5% would be three times as good. This is not true. Secondly, make sure the chemical you are using is effective for the particular pest you want to treat. While there are some broad spectrum insecticides, others are very effective only for certain pests. Also make sure that the insecticide you are using won't kill valuable plants in your home or garden.

There are three levels of toxicity indicated on labels. In this case, the toxicity these labels address is their toxic effect on humans who might be accidentally exposed. These labels are:

1. DANGEROUS—indicates very toxic compounds;
2. WARNING—indicates compounds are mildly toxic;
3. CAUTION—generally safe, only slightly toxic.

HOW IS TOXICITY DETERMINED?

Toxicity is determined in a variety of ways. Approved pesticides have been tested on experimental animals to determine how much is too much. Toxicity is always related to the amount of poison in milligrams per kilogram of body weight that will kill 50 percent of a population of experimental animals per unit of time (e.g., within two days after exposure). The animals act as models for how these poisons will affect humans in the short term. Such short-term studies are called *acute* LD/50's. Another way of measuring toxicity is the concentration of the poisons that are suspended in air or are gaseous mixtures with air. The term here is LC/50 (the lethal concentration that will kill 50 percent of the test population). Such LC/50's are based on inhaled toxicities. Some toxic labels may carry more than one LD/50, depending on the route of entry, e.g., the skin or the gastrointestinal tract. In general, the ingested LD/50 is lower, i.e., requires less poison, than dermal LD/50's.

In addition to acute toxicities, there are long-term or chronic toxic effects which may produce delayed long-term damage to the kidneys, liver, lungs or brain. Such toxins also may cause birth defects or cancer. Repeated exposures to small amounts of toxins may produce such chronic or long-term delayed effects. These too are tested on experimental animals. A number of products that accumulate in the environment and produce chronic effects are restricted or banned by state or federal regulations. Unfortunately, these were some of the most effective pesticides ever developed: DDT, dieldrin, chlordane and aldrin.

ACCIDENTAL PESTICIDE POISONING

MALATHION: An example of pesticide poisoning

Malathion is one of the more effective and widely used insecticides. Although of relatively low toxicity, it has been the cause of accidental poisonings and has been used in suicide attempts. Currently available commercial preparations—liquids, wettable powders and dusts—contain varying concentrations of malathion, which is a nervous system poison. Malathion can be absorbed from the gastrointestinal tract or lungs and even penetrate the skin. As little as four grams in a child or 25 grams in an adolescent or adult can cause serious illness or death. The time of onset of symptoms is variable, but they usually start within two hours. These symptoms include bronchial constriction, increased mucous secretion and wheezing (73% of cases), nausea and vomiting (33%), diarrhea, abdominal cramps and constipation (40%). Other symptoms may include profuse sweating, slavering (salivation) and lacrimation (shedding tears), as well as kidney and cardiovascular changes. The victim may be restless, have tremors or seizures (24% of cases) and exhibit confusion, drowsiness, slurred speech and coma (90%). Most victims recovered within 48 hours, but some deaths following malathion ingestion are well documented.

Other pesticides may be more or less toxic and present other symptoms. It is beyond the scope of this small book to go into the details of symptoms of each toxin. However, there are some rules and guidelines that *must* be followed whenever accidental poisonings occur. These guidelines are based on protocols of the Massachusetts Poison Control System and the recommendations of its Director, Dr. Frederic Lovejoy, Jr., M.D.

WHAT TO DO IN CASE OF ACCIDENTAL PESTICIDE POISONING

THE TELEPHONE NUMBER
OF POISON CONTROL IS_____
(write this in.)

Instantly call your local poison control center. Give them the following information:

1. The name of the poison (both the trade and scientific names on the container).
2. The approximate amount ingested and time of ingestion.
3. The age and approximate weight of the victim.
4. Describe as accurately as possible any symptoms you can recognize.

Do not panic! Most intoxications are not lethal, and your panic may confuse the poison control counselor as well as upset the victim.

Poison Control Centers are advisory. They will tell you what to do. If in their consideration the ingestion warrants treatment, they will probably tell you two things: First, if the victim hasn't vomited and is alert, to induce vomiting, thus emptying the stomach contents, including much of the poison (see following section on How to Induce Vomiting). Never induce vomiting if the victim is drowsy. The second thing they will tell you is to bring the victim to the local hospital emergency room. These facilities are equipped to monitor and treat patients appropriately. They will also get advice from the poison control center.

How to Induce Vomiting

If you have young children in the house, it might be a wise precaution to keep syrup of ipecac available for such emergencies. But remember: *all* potent medications should be kept out of reach (or locked away) to prevent poisoning. Adult victims who are fully conscious and alert should be given 30cc of ipecac; this can be repeated *once* in about 20 minutes. Children should get 15cc; this too can be repeated *once* in 20 minutes. *Never induce vomiting in a poisoning victim who shows signs of drowsiness or incoherence.*

What Will Happen at the Emergency Room?

The most important step in the treatment of poison ingestion is to decrease the amount of poison. At the emergency room they may wash out the stomach contents (gastric lavage) and then administer a slurry of activated charcoal, which will absorb much of the ingested toxin. They may also give a laxative, to flush the toxin through the gastrointestinal tract as rapidly as possible. They will monitor the victim's symptoms and physiologic functions and determine the appropriate treatment. Most poisonings are short-term events (up to two days), but they should be treated *only* by competent medical personnel with advice from the specialists at the poison control center.

Disposing of Pesticides

All pesticides, even those listed as relatively safe, are poisons and should be disposed of with care. There are usually explicit instructions on the label detailing how that poison should be disposed of. Sometimes labels fall off, in which case you should call your local agricultural agent or EPA office for instructions. There are some very clear guidelines that should be observed:

1. Never incinerate pesticide containers, since vaporized poisons are extremely potent and readily absorbed into the body. Spray cans are particularly dangerous, since they may explode when heated.
2. Never put any pesticide down your drain.
3. Bottle containers should be rinsed outdoors, wrapped in paper and then put into your trash.
4. Poisons containing toxic metals such as arsenic should be disposed of at an approved landfill site.
5. Always rinse (outdoors) containers used for mixing pesticides. Then wrap them in paper and put them into your trash.

PESTICIDE STORAGE AND SHELF LIFE

Because shelf life is so difficult to predict, most chemical companies recommend storing pesticides no longer than two years and, in fact, most companies will not back their products any longer than this. Your best bet to avoid the problem is to follow these recommendations:

1. Keep an accurate inventory of your stored chemicals. Rotate stock; use older materials first. It is to your benefit to use up the pesticides that you've purchased as soon as possible. Don't put them in the back room and forget about them.
2. Store pesticides in a dry, well-ventilated place at temperatures above freezing.
3. Always keep a pesticide in its original container and make sure it is tightly sealed.
4. Store granular or powdered materials above ground to avoid dampness.
5. Keep the temperature under 100°F if storing volatile compounds.
6. Keep volatile herbicides separate from other pesticides to avoid contamination.
7. Keep liquids out of sunlight to avoid photo-decomposition.
8. Consult the label for special storage instructions.

How to Store Your Pesticides

1. Partition off a closet or section of a room that can be locked and ventilated and use this for storage.
2. The storage area should be kept cool and dry and should be well ventilated. This will prolong the useful life of your pesticides.
3. The storage area must be kept *locked* when not in actual use.
4. Durable warning signs such as DANGER: PESTICIDE STORAGE AREA—KEEP OUT should be placed on all doors and windows.
5. Always have a supply of absorptive clay, vermiculite or kitty litter on hand to soak up spills. Also have a shovel, broom and dust pan. These should not be used for any other purposes.
6. *Never* store food in the storage area.
7. Tell your local fire department the location of your pesticide storage area and let them know the types and quantities of materials you ordinarily have on hand.
8. Always keep pesticides in their original containers. Make sure the labels are attached and the containers are clean, not

rusted, tightly sealed and up off the floor. Place torn bags inside clear plastic bags—do not re-bag into unlabeled bags.

9. Never allow the temperature in the storage area to get above 100°F or drop below freezing.
10. Keep your pesticide storage area clean, organized and in good repair. This will help you to reduce accidents.

PESTICIDE DISPOSAL FOR THE HOMEOWNER

A commonly asked question is how the homeowner should dispose of unwanted pesticides. These are usually small containers (quarts, one-pound bags, etc.) of various commonly used insecticides such as malathion, diazinon, carbaryl (Sevin), etc. or fungicides such as captan, sulfur, etc. This section covers how homeowners should handle this problem.

METHODS FOR THE HOMEOWNER

The first rule is to buy only what you need and to store it carefully so you can use it up. This eliminates most future problems. However, you may move into a house or apartment and find assorted pesticides no longer in use, accumulated in the past by someone else. How can these be disposed of?

1. Send the material to your local landfill, a few containers at a time.
2. Dilute with water and dump in shallow pits on your own property.
3. Bury, without diluting, on your own property.

Option No. 1: Send to a landfill

This is a good way to go, but be careful not to send too much at once. Wrap all pesticide containers in several layers of newspaper so they won't break in transit and endanger the trash collectors. For urban and dense suburban situations, this is the only practical disposal method.

Option No. 2: Dilute and bury at home

For urban, suburban or rural locations, some pesticides can be disposed of on your own property. This can be done most safely by *diluting* the pesticide before disposal. For example, a good way to dispose of a half-full bottle of unwanted garden insect spray (say, a 50% malathion product) would be to mix this with about five gallons of water and then pour this diluted material into a shallow trench *in a suitable location*. Avoid *sandy areas,* areas of *bedrock* or steep slopes where run-off is likely, areas near water (springs, brooks, ponds) or areas used for gardens or crops. Good locations would be relatively flat, well-drained wood lots or fallow fields where there is enough soil to absorb the liquid. Always cover the spot with clean soil when you are done.

Option No. 3: Bury at home without dilution

To bury pesticides, select a piece of your property that is suitable (see above) and that *will not be disturbed* by construction or any other excavation for at least five years, preferably longer. Dig a hole about 18 to 20 inches deep and put two to three inches of crushed charcoal at the bottom. Next, put down two to three inches of lime. Then pour or dump in the pesticide. Do not put in sealed containers. To break down, pesticides must make contact with the soil so that bacteria can act on the chemicals. Cover with at least 12 inches of clean soil.

Local landfill and home disposal of pesticides are undesirable, but necessary. These methods should *not* be relied on for future needs (i.e., you should plan ahead and use up pesticide supplies according to their labeled uses) but are needed to cope with existing problems accumulated in the past. These methods should *not* be used for highly toxic pesticides with the skull and crossbones on the label nor for large quantities (more than a few bottles). Be careful not to pollute water or arable soil while disposing of pesticides at home. Do not bury unopened containers.

Index